Zhaodong Wang
**Sino-British Negotiations and the Search for a Post-War Settlement, 1942–1949**

# Transformations of Modern China

Edited by
Jennifer Altehenger, Daniel Leese, Nicola Spakowski,
and Sebastian Veg

## Volume 3

Zhaodong Wang

# Sino-British Negotiations and the Search for a Post-War Settlement, 1942–1949

Treaties, Hong Kong, and Tibet

**DE GRUYTER**
OLDENBOURG

ISBN 978-3-11-070642-0
e-ISBN (PDF) 978-3-11-070665-9
e-ISBN (EPUB) 978-3-11-070671-0
ISSN 2511-6029

**Library of Congress Control Number: 2021940492**

**Bibliographic information published by the Deutsche Nationalbibliothek**
The Deutsche Nationalbibliothek lists this publication in the Deutsche Nationalbibliografie;
detailed bibliographic data are available on the Internet at http://dnb.dnb.de.

© 2022 Walter de Gruyter GmbH, Berlin/Boston
Cover image: Ceremony at the Cenotaph at Central, Hong Kong marking the liberation of
Hong Kong, 30 Aug 1945, The World War II Database by C. Peter Chen.
Printing and binding: CPI books GmbH, Leck

www.degruyter.com

MIX
Papier aus verantwor-
tungsvollen Quellen
FSC® C083411

# Acknowledgements

The book began as a doctoral dissertation. It would not have come to fruition without help from many hands. First and foremost, I must express my thanks to my family, particularly my wife, Wuyao Liu. Without their understanding and support, this book would not have been started and completed.

I am extremely grateful to my supervisors at University of Edinburgh – Dr. Felix Boecking and Dr., now Professor, Jeremy Crang – for their excellent and cooperative supervision. The thesis has benefited significantly from their expertise and professionalism in their respective fields of Chinese and British history. This has encouraged me to seek to balance my discussion between Chinese and British perspectives. Dr. Crang has also offered me enormous help and advice in relation to the thesis structure and my written English.

I also owe a debt of gratitude to the Collaborative Innovation Centre for Territorial Sovereignty and Maritime Rights at Wuhan University, and to the Chinese Scholarship Council. These institutions funded my doctoral studies at Edinburgh and allowed me to focus my time and energy on this research. My gratitude also extends to Professor Dekun Hu at Wuhan University who seven years ago re-ignited my passion for study abroad. The initial idea for this book topic is derived from him.

This thesis incorporates significant archival research. I am indebted to the Simon Fennell Award, and to Research Student Support Fund, in the School of History, Classics and Archaeology at Edinburgh which helped finance my two visits to London to work in the National Archives and the British Library. Further thanks are due to Dr. Xuelei Huang in the School of Literatures, Languages and Cultures at Edinburgh, and to Professor Li Zhang at the Institute of Modern History, Academia Sinica, Taipei. They helped to facilitate my research trip to Taipei, for which I am very grateful.

I want to express thanks to Dr. Francesca Kaufman, now on the academic staff at University of Manchester, who proofread most of my original drafts and provided valuable suggestions and criticisms. My thanks also extend to two examiners of my PhD dissertation, Professor Alvin Jackson at University of Edinburgh and Dr. Tehyun Ma at University of Sheffield, and two anonymous reviewers of my book manuscript. Their comments and revision suggestions are valuable to this book.

Last but not least, I want to express my gratitude to book editors of De Gruyter Publishers, Rabea Rittgerodt, Jana Fritsche and Antonia Mittelbach. Thanks for their hard work and support; without them the book would not be published in such a high-quality fashion.

Zhaodong Wang
Beijing, 2021

# Abstract

Despite the vast research by scholars on international history during the era of the Pacific War, comparatively little has been written about the bilateral relationship between the Chinese Nationalist government and the British government and their discussions during 1942–1949 over a post-war settlement in Asia. These were dominated by two underlying themes: the elimination of the British imperialist position in China and the establishment of an equal and reciprocal bilateral relationship. In particular, these discussions focused on three matters: treaties (the 1943 Sino-British treaty and the discarded Sino-British commercial treaty); the future of Hong Kong; and the political status of Tibet. Drawing on archival sources in Britain, the United States and China, the book demonstrates that negotiations over a post-war Sino-British settlement had an encouraging start in 1942–43 but by 1949 had failed to reach a satisfactory settlement. Not only did they fail to rebuild the two countries' commercial relations on an equal and reciprocal basis but they also did not terminate the informal British empire in China. The reasons for the failure were complex, encompassing both internal and external factors, including the powerful influence of the United States.

https://doi.org/10.1515/9783110706659-002

# Contents

# Acronyms and Abbreviations

| | |
|---|---|
| AH | Academia Historica, Taipei and Hsin-tien |
| BAAG | British Army Aiding Group |
| BDFA | British Documents on Foreign Affairs |
| BO | Burma Office |
| BT | Board of Trade |
| CAB | Cabinet Office |
| CASS | Chinese Academy of Social Sciences, Beijing |
| CCP | Chinese Communist Party |
| CHAR | Churchill Archive Records |
| CMCS | Chinese Maritime Customs Service |
| CNP | Chinese Nationalist Party |
| CO | Colonial Office |
| DBPO | Documents on British Policy Overseas |
| DO | Dominion Office |
| FRUS | Documents of Foreign Relations of the United States |
| FO | Foreign Office |
| GRC | Guomindang Revolutionary Committee |
| HKPU | Hong Kong Planning Unit |
| HO | Home Office |
| IMH | Institute of Modern History, Academia Sinica, Taipei |
| IO | India Office |
| IOR | India Office Records |
| ISBMAC | Investigation and Statistics Bureau of the Military Affairs Commission (of the Nationalist Government) |
| MAC | Military Affairs Commission (of the Nationalist Government) |
| MFN | Most Favoured Nation |
| ML | Ministry of Labour |
| MT | Ministry of Transport |
| MTAC | Mongolian and Tibetan Affairs Commission (of the Executive Yuan) |
| OCAC | Overseas Chinese Affairs Commission (of the Executive Yuan) |
| OSS | Office of Strategic Services, the United States |
| *SLCB* | *Zhonghuaminguo zhongyao shiliao chubian (Kangrizhanzheng shiqi): zhanshi waijiao* [Preliminary Compilation of Important Historical Sources Relating to the Republic of China (the Period of the Anti-Japanese War): Wartime Diplomacy] |

https://doi.org/10.1515/9783110706659-003

SLGB    *Jiang Zhongzheng zongtong dang'an: shilüe gaoben* [Draft Papers of
        Chiang Kai-shek]
SNDCG   Supreme National Defence Council of the Guomindang
SOE     Special Operations Executive
TBL     The British Library, London
TFB     Tibetan Foreign Bureau
TNA     The National Archives, Kew, London
TRP     Tibetan Revolutionary Party
UNESCO  United Nations Educational, Scientific and Cultural Organisation
UNECAFE United Nations Economic Commission for Asia and the Far East
WO      War Office

# List of Maps

https://doi.org/10.1515/9783110706659-004

# Note on the Text

In this book, the term Chinese Nationalist government (1927–1949) is used inter-changeably with that of the Chinese government. Similarly, Guomindang 國民黨 (Kuomintang) is used interchangeably with the Chinese Nationalist Party. The re-gime that the Chinese Communists established is referred to as the Chinese Com-munist government or regime. The term "East Asia" in the book is used to refer to a wider geographical area covering East and Southeast Asia, and North Asia, which has basically the same meaning as the Europe-centred term "the Far East".

https://doi.org/10.1515/9783110706659-005

# Romanisation

*Pinyin* romanisation is used for Chinese names and terms, except for those of Chiang Kai-shek and Sun Yat-sen, as this style is more widely recognised than their *pinyin* names: Jiang Jieshi 蔣介石 and Sun Zhongshan 孫中山. In general, Chinese characters will follow their *pinyin* names or terms immediately when they are referred to in the text for the first time. In some cases, names or terms romanised in other ways are additionally provided in brackets because they were widely used during the time period that the book concerns. With regard to Tibetan names, I adopt a style that is commonly used in English language works. For the sake of readers with a Chinese-language background, *pinyin* terms and Chinese characters are supplemented in brackets following their corresponding Tibetan names when they occur in the text for the first time.

https://doi.org/10.1515/9783110706659-006

# Introduction

## Overview

> The day of August 29 [1942] is the 100th anniversary of the Nanjing Treaty and thus tremendously significant. If it can be capitalised on properly, the significance will be even greater. To announce on the day the abolition of 'unequal treaties' can not only accomplish a major goal of the [Guomindang's] revolution but also comfort the soul of the Premier Sun Yat-sen in Heaven. Equality must be fought for by ourselves but not others, and the attainment of equality should not be placed at the mercy of others' agreement. Different from India's begging for its freedom, we should follow the example of Mustafa Kemal Atatürk and declare the abolition in revolutionary spirit and by quick and resolute means. At the same time, the Chinese public should be told to befriend other nations in the spirit of friendship, in the manner of a great nation and by no way of either obsequiousness or arrogance. The proclamation of terminating unequal treaties will add great glory to the [Guomindang's] revolution history and let it gain more admiration and the support of the whole nation.[1]

These words are quoted from a letter from Lin Yutang 林語堂, a famous Chinese writer and scholar who then lived in Washington, to Chiang Kai-shek. He suggested that Chiang terminate unequal treaties in China on the hundredth anniversary of the Nanjing Treaty. Chiang, nevertheless, hesitated and decided to first consult his advisors on the Committee of Discussing International Questions, a temporary body affiliated to the Supreme National Defence Council of the Guomindang (SNDCG), in which the power of the Chinese government was concentrated during wartime. Chiang's reluctance can be easily understood and the committee's conclusion offers some insight: it was an inappropriate time to adopt an abolition approach given that the war was still in progress and Britain and the United States had already committed themselves to cancellation of their treaty rights after the war ended.[2] To the Chinese Nationalists, war emergencies apparently far outweighed other priorities, such as suppressing Chinese Communism or confronting western imperialism. However, this was not al-

---

1 CNP Archives, 防005/0007, *Guojiwenti taolunhui fengling yanjiu jie Nanjing tiaoyue bainian zhengshi xuangao feizhi bupingdeng tiaoyue zhi jianyi* [Advice given by the Committee of Discussing International Questions about an official announcement abolishing unequal treaties by exploiting the 100th anniversary of the Nanjing Treaty], letter from Chiang Kai-shek to Wang Chonghui and Wang Shijie, August 3, 1942.
2 CNP Archives, 防005/0007, *Guojiwenti taolunhui fengling yanjiu jie Nanjing tiaoyue bainian zhengshi xuangao feizhi bupingdeng tiaoyue zhi jianyi* [Advice given by the Committee of Discussing International Questions about an official announcement abolishing unequal treaties by exploiting the 100th anniversary of the Nanjing Treaty], minute by the Secretariat of the SNDCG, August 1942.

https://doi.org/10.1515/9783110706659-007

ways the case. Only two months later, Chiang's mind changed significantly. This can be attributed to the fact that Britain and the United States agreed to an immediate cancellation of extraterritoriality and other related treaty rights in China. He was inspired to pursue a complete and swift end to foreign imperialism in China, which in the case of Britain took the treaty privileges beyond the scope of extraterritoriality and those regarding Hong Kong and Tibet into consideration. This strained the already tense Sino-British alliance and resulted in an earlier start to discussions between the Chinese Nationalist government and the British government over their post-war relations. From then on, the discussions were deeply intertwined with war developments and the Allied nations' plans for a post-war world settlement, leading to mixed results both from Chinese and British perspectives.

The subject of this book are these discussions during the period from 1942 to 1949, covering the war and the years immediately following. They are dominated by two underlying themes: the elimination of the British imperialist position in China and the establishment of an equal and reciprocal bilateral relationship. In particular, the discussions focused on three matters: treaties (the 1943 Sino-British treaty and the eventually discarded Sino-British commercial treaty), the future of Hong Kong and the political status of Tibet. As it will demonstrate, China's Second World War was not just a war against Japan but also a struggle to end western imperialism, in spite of the fact that China's wartime alliance with Britain and the United States at times conflicted with this aim. The book is also a study on the reasons for the protracted and incomplete retreat of British imperialist interests in China by 1949. Despite a promising start in 1942, Sino-British negotiations over a post-war settlement during this period eventually failed both to eradicate British imperialism in China and to fully establish their post-war relations on the basis of equality and reciprocity.

## The Allied Nations' War Strategies and Post-war Plans for East Asia (the Far East)

As a backdrop to the investigation of Sino-British negotiations between 1942 and 1949, it is first necessary to look at the broader strategic picture in the region. Although united in defeating the Japanese Empire and restoring peace in the region, the four major Allied countries – Britain, China, the United States and the Soviet Union – had different wartime priorities and post-war blueprints for East Asia.

Britain and the United States advocated a "Europe first" strategy which, simply put, entailed the defeat of Nazi Germany before concentrating military forces against the Japanese Empire.[3] Understandably, this perturbed the Chinese government. From its perspective, this strategy indicated that only limited western military resources would be available in the Pacific and China theatre. After four years of fighting alone against the Japanese, Chongqing was on the brink of collapse and urgently needed to have the Burma Road reopened, the main overland transport route through which China could access external supplies which had been cut off by the Japanese seizure of British Burma in 1942. Yet until Germany was defeated, Britain's primary concern in East and South Asia was to defend India rather than regain Burma. The United States was keener to support China's war effort militarily, but the US focus was on its naval offensive in the Pacific and it was concluded by army planners in early 1944 that "by the time the Chinese Army was adequately trained and equipped it would be too late to be of any assistance in the fight against Japan."[4] Meanwhile, Washington invited the Soviet Union to shoulder more of the responsibility for the war in the Pacific.[5] Moscow, however, was in no mood for aggressive military action against Japan in the region while the war in Europe was still in progress and it was engaged in a life and struggle for its very existence. It was not until August 1945 that the USSR finally joined the war and attacked Japanese forces in Northeast China (Manchuria).

As for post-war arrangements in East Asia, views of the main players were similarly divergent. In broad terms, the aspirations of the Americans were laid down in the Atlantic Charter of 1941 which promised "the right of all peoples to choose the form of government under which they will live" and "the enjoyment of all states ... of access, on equal terms, to the trade and to the raw materials which are needed for their economic prosperity."[6] China also hoped for a post-war world exempt from European colonialism but its focus was more on how this would play out in its own sphere. Besides recovering territories occupied by the Japanese since 1894, Chongqing desired for a full end to foreign imperialism in China, particularly in relation to the imperialism of Britain, and a rebuilding of its relations with foreign powers on an equal footing.

Yet British thinking was at variance with these Sino-American desires. London's war aim, particularly in East Asia, was largely the preservation of British

---

3 Kori Schake, *Safe Passage: the Transition from British to American Hegemony* (Cambridge, Massachusetts: Harvard University Press, 2017), 264–266.
4 Christopher Thorne, *Allies of a Kind: the United States, Britain, and the War against Japan, 1941–1945* (New York: Oxford University Press, 1978), 428.
5 Ibid., 497–498.
6 Quoted in ibid., 101–102.

colonies and imperial preference in trading terms.[7] This meant a resumption of British authority over Burma, Malaya, Singapore, North Borneo and Hong Kong, once the Japanese forces were expelled. To a large extent, Britain's agreement to the Atlantic Charter was thus an expedient to bring the United States into alignment with the British war effort.[8] Whitehall was also opposed to any US sponsorship of China as a great power and the inclusion of China in international discussions over the future of East Asia. Nevertheless, some concessions were made. In January 1943, the Americans and British abolished their extraterritorial and other related rights in China. Likewise, matters relating to the future of China were brought to the Allied Cairo Conference in November of that year for resolution: the recovery of Northeast China (occupied by Japan since 1931), Taiwan and the Pescadores (occupied by Japan since 1895) and the independence of Korea (a traditional protectorate of the Chinese Empire but annexed by Japan in 1910).[9]

The year 1944, however, witnessed some significant changes in Sino-American and Anglo-American relations. Washington became increasingly impatient at the perceived low morale and poor military performance of the Chinese forces, coupled with alleged corruption inside the Nationalist regime and its incessant hostilities with the Chinese Communist Party (CCP). In addition, the personal antipathy between Chiang Kai-shek and General Joseph Stilwell, Chiang's US Chief of Staff in the China Theatre and the supreme commander of US forces in China, India and Burma (1942–1944), further alienated President Franklin Roosevelt from the Nationalist authorities.[10] Thereafter, China was, intentionally or unintentionally, left absent from some of the key conferences in determining East Asia's fate.[11] For example, at the Second Anglo-American Quebec Conference in September 1944, it was agreed, without the participation of the Chinese, that the British would wage a campaign in the direction of Rangoon and Singapore.[12] Meanwhile, although friction still existed between them, the Anglo-American tension over colonial issues was relaxed to some degree. The two countries

---

**7** Schake, *Safe Passage*, 266.

**8** Ibid., 263.

**9** Xiaoyuan Liu, *A Partnership for Disorder: China, the United States, and Their Policies for the Postwar Disposition of the Japanese Empire, 1941–1945* (New York: Cambridge University Press, 1996), 131–144.

**10** Thorne, *Allies of a Kind*, 429–431.

**11** It is important to note that China continued actively participating in the establishment of a series of international bodies, such as the United Nations, the World Bank and the International Monetary Fund.

**12** Thorne, *Allies of a Kind*, 416.

cooperated at the San Francisco Conference in April-May 1945 in laying down the goal of "self-government" for colonial territories in the Charter of the United Nations while dispensing with the term "independence" as desired by the Chinese and others.[13]

As for the Soviet Union, it was drawn into the post-war arrangements for East Asia towards the end of the war. Its aim was clear, namely the restoration of Russian imperial interests lost after the Russo-Japanese War of 1904–05. At the Yalta Conference in February 1945, Roosevelt agreed to these demands in order to secure an earlier Soviet participation in the war against Japan. Yet those concessions which impacted on China's interests, such as the independence of Outer Mongolia (a former part of the Qing Empire) and the Soviet Union's exclusive and special position in relation to the two warm-water ports and two railway lines in Northeast China,[14] were made without reference to the Chinese government.[15] Nevertheless, the Nationalist government eventually accepted these terms, mainly out of fear of the Soviet Russia's support for the CCP.

## The Sino-British Context and the Aims of the Book

Before the Pacific War, the British were on the back foot in China. Since the 1920s, Chinese nationalism was on the rise and its proponents demanded the full abolition of the "unequal treaties" that had been imposed on China by the Western powers and Japan since the First Opium War (1839–1842). Nonetheless, despite being under pressure in the region, Britain's "informal empire"[16] in China was still much in evidence before the war. It included the treaty-port system, two British concessions in Tianjin 天津 (Tientsin) and Guangzhou 廣州 (Canton), two British-dominated International Settlements in Shanghai 上海 and Gulangyu 鼓浪嶼 (Kulangsu), the British legation quarter in Beijing, the

---

13 Ibid., 599.
14 The two ports were the naval port of Lüshun 旅順 (Port Arthur) and the commercial port of Dalian 大連 (Dairen); the two railways were the Chinese Eastern Railway from Manzhouli 滿洲里 (Manchuli) to Suifenhe 綏芬河 and the South Manchurian Railway from Ha'erbin 哈爾濱 (Harbin) to Dalian and Lüshun. Through these lines, Far East Russia was connected to Dalian and Lüshun.
15 Liu, *A Partnership for Disorder*, 242–243.
16 "Informal empire" is a term to describe strong British influence in trade and commerce, and degrees of informal political control, over territories beyond the British Empire. It mainly applies to places such as China, the Middle East and Latin America. See William Roger Louis, "Foreword," in *The Oxford History of the British Empire: Historiography*, 5 vols., vol. 5, ed. Robin W. Winks (Oxford: Oxford University Press, 1999), ix–x.

British Crown Colony of Hong Kong and a British-dominated Chinese Maritime Customs Service (CMCS). Furthermore, the British retained many "treaty rights" in China, principally extraterritoriality, and enjoyed a special position in Tibet, supporting the latter's autonomy against Chinese intervention and controlling its trade with the outside. Yet the war in many ways catalysed the decolonisation process. First, the Japanese occupied all British territories, leased or ceded, in China, as well as the CMCS. Although a rival service headquarters was subsequently set up in Chongqing, it had lost its control of Japanese-occupied China, where most Chinese tariff revenue was generated. Second, the Sino-British military alliance made Britain's imperialist position in China seem incongruous.

The Pacific War drove China and Britain into the same camp against the Japanese Empire, but disputes over the "informal empire" still plagued their wartime relationship. The two countries frequently clashed over three major questions: how to terminate the old and unequal Sino-British treaties and rebuild their post-war relations on an equal and reciprocal basis; whether, or in what way, Hong Kong would be returned to China after the war; and whether the British would recognise Chinese sovereignty, rather than suzerainty, over Tibet.

Discussions over the Sino-British post-war settlement had a promising start but an unsatisfactory end. In January 1943, Britain terminated its extraterritoriality and the treaty-port system. Several important questions were, however, left unresolved. A Sino-British commercial treaty failed to materialise (whereas a Sino-American counterpart was agreed in 1946); Hong Kong retained its colonial status until 1997; and British interests in Tibet, as well as those in the disputed McMahon area, were inherited by independent India in 1947 without hindrance. In this sense, the Pacific War did not eradicate British imperialism in China. Moreover, these issues, which might have had a chance of being resolved in tandem with the global post-war arrangements triggered by the Pacific War, were eventually dragged into the Cold War, when a different Chinese regime on the mainland assumed power.

Drawing on detailed case studies of these three important areas of contention – treaties, Hong Kong and Tibet – and on both British and Chinese sources, this study attempts to address how and why the British and Chinese governments failed to arrive at a satisfactory post-war settlement before the change of regime in China 1949. It is also hoped that some wider conclusions might, in turn, be drawn about the post-war international order in East Asia and the Asian Pacific. It is the author's contention that it is over-simplistic to address the issue of the failure of the Sino-British deliberations merely through references to the stubbornness, or weaknesses, of each party, or the disturbance of the

Chinese civil war. Behind the protracted negotiations lay a more complex set of domestic considerations and external influences.

## Literature Review

As far as the author is aware, no scholarly work has systematically examined Sino-British discussions in the 1940s in relation to their post-war relationship, or indeed has conjointly analysed the three Sino-British areas of contention under consideration here (the treaties, Hong Kong and Tibet) and placed them in the context of wider Allied discussions over post-war global arrangements. Nonetheless, this study draws upon a rich array of historiography in four relevant fields: studies of Britain's policy towards China, Chinese nationalism, British imperialism and the Pacific War.

Turning first to specialist studies of Britain's policy towards China in the 1940s, research in this field has tended to focus on British economic interests in China and Hong Kong. Key works in this area include Aron Shai's *Britain and China, 1941–47: Imperial Momentum*[17] and Feng Zhongping's *The British Government's China Policy, 1945–1950*.[18] Meanwhile, other volumes in this field have more of a Chinese-Communist focus. Among these are Brian Porter's *Britain and the Rise of Communist China: A Study of British Attitudes, 1945–1954*[19] and James Tuck-Hong Tang's *Britain's Encounter with Revolutionary China, 1949–54*.[20] There are also more general studies such as Li Shi'an's *Taipingyangzhanzheng shiqi de Zhong-Ying guanxi* (The Sino-British Relationship during the Pacific War).[21]

There are also relevant studies that relate to the history of Chinese nationalism. These include Rana Mitter's *A Bitter Revolution: China's Struggle with the*

---

17 Aron Shai, *Britain and China, 1941–47: Imperial Momentum* (London: Macmillan in Press association with St. Antony's College, Oxford, 1984).

18 Zhong-ping Feng, *The British Government's China Policy, 1945–1950* (Keele: Keele University Press, 1994).

19 Brian Porter, *Britain and the Rise of Communist China: A Study of British Attitudes, 1945–1954* (London and Toronto: Oxford University Press, 1967).

20 James Tuck-Hong Tang, *Britain's Encounter with Revolutionary China, 1949–54* (London and New York: St. Martin's Press, 1992).

21 Li Shi'an, *Taipingyangzhanzheng shiqi de Zhong-Ying guanxi [The Sino-British Relationship during the Pacific War]* (Beijing: China Social Sciences Press, 1994). The latest version is ibid., *Zhanshi Yingguo dui Hua zhengce [Britain's China Policy during the Second World War]*, ed. Dekun Hu, *Fanfaxisi zhanzheng shiqi de Zhongguo yu Shijie yanjiu [Research on China and the World during the Anti-Fascist War]*, 9 vols., vol. 7 (Wuhan: Wuhan University Press, 2010).

*Modern World*,[22] Hans van de Ven's *War and Nationalism in China, 1925 – 1945*[23] and Robert Bickers's *Out of China: How the Chinese Ended the Era of Western Domination*.[24] Alongside these works, there are specialist Chinese studies of the liquidation of the "unequal treaties." These include Li Yumin's *Zhongguo feiyue shi* (The History of Abolishing Unequal Treaties in China)[25] and Wang Jianlang's *Zhongguo feichu Bupingdeng tiaoyue de licheng* (Historical Process of Chinese Abolition of Unequal Treaties).[26] In tandem with these sources are apposite works on British imperialism in China (and East Asia or the Far East more generally). These encompass Edmund S.K. Fung's *The Diplomacy of Imperial Retreat: Britain's South China Policy, 1924 – 1931*,[27] Irving Friedman's *British Relations with China, 1931 – 1939*,[28] Nicholas Clifford's *Retreat from China: British Policy in the Far East, 1937 – 1941*,[29] Peter Lowe's *Britain in the Far East: A Survey from 1819 to the Present*[30] and Robert Bickers' *Britain in China: Community, Culture and Colonialism 1900 – 1949*.[31]

There are a number of pertinent studies of the Pacific War. Among these are works on China's role in the conflict such as Rana Mitter's *China's War with Japan 1937 – 1945: the Struggle for Survival*[32] and Hans van de Ven's *China at War: Tri-*

---

**22** Rana Mitter, *A Bitter Revolution: China's Struggle with the Modern World* (New York: Oxford University Press, 2005).

**23** Hans van de Ven, *War and Nationalism in China, 1925 – 1945* (London and New York: Routledge Curzon, 2003).

**24** Robert Bickers, *Out of China: How the Chinese Ended the Era of Western Domination* (London: Allen Lane – Penguin Random House, 2017).

**25** Yumin Li, *Zhongguo feiyue shi [The History of Abolishing Unequal Treaties in China]* (Beijing: Zhonghua Shuju, 2005).

**26** Jianlang Wang, *Zhongguo feichu Bupingdeng tiaoyue de licheng [The Historical Process of Chinese Abolition of Unequal Treaties]* (Nanchang: Jiangxi Renmin Chubanshe, 2000).

**27** Edmund S.K. Fung, *The Diplomacy of Imperial Retreat: Britain's South China Policy, 1924 – 1931* (Hong Kong: Oxford University Press, 1991).

**28** Irving S. Friedman, *British Relations with China, 1931 – 1939* (New York and London: Institute of Pacific Relations and Allen & Unwin, 1940).

**29** Nicholas Clifford, *Retreat from China: British Policy in the Far East, 1937 – 1941* (Seattle: University of Washington Press, 1967).

**30** Peter Lowe, *Britain in the Far East: A Survey from 1819 to the Present* (London: Longman, 1981). Despite the fact that the title comprises the entirety of the "Far East", China actually remains a major concern for the author.

**31** Robert Bickers, *Britain in China: Community, Culture and Colonialism 1900 – 1949* (Manchester and New York: Manchester University Press, 1999).

**32** Rana Mitter, *China's War with Japan 1937 – 1945: the Struggle for Survival* (London: Penguin Books, 2013).

*umph and Tragedy in the Emergence of the New China, 1937–1952.*[33] There are also studies of the Anglo-American relationship during the war – Christopher Thorne's *Allies of a Kind: the United States, Britain, and the War against Japan, 1941–1945*[34] and William Roger Louis' *Imperialism at Bay: the United States and Decolonization of the British Empire, 1941–1945*[35] – and of their relationship in the region during the immediate post-war period: Christopher Baxter's *The Great Power Struggle in East Asia, 1944–50: Britain, America and Post-war Rivalry*[36] and Xiang Lanxin's *Recasting the Imperial Far East: Britain and America in China, 1945–1950.*[37] Meanwhile, the Sino-American relationship is covered by Ch'i Hsi-sheng's *The Much Troubled Alliance: US-China Military Cooperation during the Pacific War, 1941–1945*[38] and Liu Xiaoyuan's *A Partnership for Disorder: China, the United States, and Their Policies for the Postwar Disposition of the Japanese Empire, 1941–1945,*[39] and aspects of the Soviet-British-American perspective feature in David Reynolds and Vladimir Pechatnov's edited book, *The Kremlin Letters: Stalin's Wartime Correspondence with Churchill and Roosevelt.*[40]

To add to these sources are specialist studies of specific topics relevant to the Sino-British relationship. These include works relating to extraterritoriality, such as Wesley Fishel's *The End of Extraterritoriality in China* and Emily Whewell's "British Extraterritoriality in China: the Legal System, Functions of Criminal Jurisdiction, and Its Challenges, 1833–1943"[41]; those concerned with the Chi-

---

[33] Hans van de Ven, *China at War: Triumph and Tragedy in the Emergence of the New China, 1937–1952* (London: Profile Books, 2017).

[34] *Thorne, Allies of a Kind.*

[35] William Roger Louis, *Imperialism at Bay: the United States and Decolonization of the British Empire, 1941–1945* (New York: Oxford University Press, 1977).

[36] Christopher Baxter, *The Great Power Struggle in East Asia, 1944–50: Britain, America and Post-war Rivalry* (London: Palgrave Macmillan, 2009).

[37] Lanxin Xiang, *Recasting the Imperial Far East: Britain and America in China, 1945–1950* (New York: M. E. Sharpe, 1995).

[38] Hsi-sheng Ch'i, *The Much Troubled Alliance: US-China Military Cooperation during the Pacific War, 1941–1945* (London: World Scientific Publishing Co., 2015).

[39] *Liu, A Partnership for Disorder.*

[40] David Reynolds and Vladimir Pechatnov, eds., *The Kremlin Letters: Stalin's Wartime Correspondence with Churchill and Roosevelt* (New Haven and London: Yale University Press, 2018).

[41] Wesley R. Fishel, *The End of Extraterritoriality in China* (Berkeley and Los Angeles: University of California Press, 1952); Emily Whewell, "British Extraterritoriality in China: the Legal System, Functions of Criminal Jurisdiction, and Its Challenges, 1833–1943," PhD diss. (University of Leicester, 2015). Other pertinent works are Thomas B. Stephens, *Order and Discipline in China: the Shanghai Mixed Court 1911–27* (Seattle and London: University of Washington Press, 1992); Turan Kayaoglu, *Legal Imperialism: Sovereignty and Extraterritoriality in Japan, the Ottoman Empire, and China* (Cambridge: Cambridge University Press, 2010); Pär Kristoffer Cassel, *Grounds of*

nese Maritime Customs Service, such as Hans van de Ven's *Breaking with the Past: the Maritime Customs Service and the Global Origins of Modernity in China* and Chen Shiqi's *Zhongguo jindai haiguan shi* (The Modern History of the Chinese Maritime Customs Service)[42]; those dealing with Hong Kong, such as Kit-Ching Lau Chan's *China, Britain and Hong Kong, 1895–1945*, Steven Tsang's *Hong Kong: An Appointment with China* and Sun Yang's *Wuguo'erzhong: Zhanhou Zhong-Ying Xianggang wenti jiaoshe, 1945–1949* (An Ending Without Result: Sino-British Post-war Discussions over the Matter of Hong Kong, 1945– 1949)[43]; and those covering Tibet, such as Alastair Lamb's *Tibet, China &*

---

*Judgement: Extraterritoriality and Imperial Power in Nineteenth-Century China and Japan* (Oxford and New York: Oxford University Press, 2012); Douglas Clark, *Gunboat Justice: British and American Law Courts in China and Japan, 1842–1943*, 3 vols., vol. 1– 3 (Hong Kong: Earnshaw Books Limited, 2015); Shi'an Li, "The Extraterritoriality Negotiations of 1943 and the New Territories," *Modern Asian Studies* 30, no. 3 (1996); Francis Clifford Jones, *Extraterritoriality in Japan and the Diplomatic Relations Resulting in Its Abolition, 1853–1899* (New Haven: Yale University Press, 1931); G.W. Keeton, *The Development of Extraterritoriality in China*, 2 vols., vol. 1– 2 (London: Longmans, Green and Company, 1928); Robert Bickers, "Legal Fiction: Extraterritoriality as an Instrument of British Power in China in the 'Long Nineteenth Century'," in *Empire in Asia: A New Global History (The Long Nineteenth Century)*, 2 vols., vol. 2, ed. Donna Brunero and Brian P. Farrell (London: Bloomsbury Academic, 2018); Francis Piggott, *Exterritoriality: the Law Relating to Consular Jurisdiction and to Residence in Oriental Countries* (London: William Clowes and Sons Limited, 1892); William Edward Hall, *A Treatise on the Foreign Powers and Jurisdiction of the British Crown* (Oxford: Clarendon Press, 1894); Charles James Tarring, *British Consular Jurisdiction in the East: With Topical Indices of Cases on Appeal from and Relating to Consular Courts and Consuls, also a Collection of Statutes Concerning Consuls* (London: Stevens and Haynes, 1887).

**42** Hans van de Ven, *Breaking with the Past: the Maritime Customs Service and the Global Origins of Modernity in China* (New York: Columbia University Press, 2014); Shiqi Chen, *Zhongguo jindai haiguan shi [The Modern History of the Chinese Maritime Customs Service]* (Beijing: Renmin Chubanshe, 2002). Other pertinent works are: Chihyun Chang, *Government, Imperialism and Nationalism in China: the Maritime Customs Service and Its Chinese Staff* (London and New York: Routledge, 2013); Robert Bickers, "Revisiting the Chinese Maritime Customs Service, 1854–1950," *The Journal of Imperial and Commonwealth History* 36, no. 2 (2008); Donna Brunero, *Britain's Imperial Cornerstone in China: the Chinese Maritime Customs Service, 1854–1949* (London and New York: Routledge, 2006); Felix Boecking, *No Great Wall: Trade, Tariffs, and Nationalism in Republican China, 1927–1945* (Cambridge (Massachusetts) and London: Harvard University Asia Center, 2017).

**43** Kit-Ching Lau Chan, *China, Britain and Hong Kong, 1895–1945* (Hong Kong: The Chinese University Press, 1990); Steve Tsang, *Hong Kong: An Appointment with China* (London and New York: I.B. Tauris & Co Ltd, 1997); Yang Sun, *Wuguo'erzhong: Zhanhou Zhong-Ying Xianggang wenti jiaoshe, 1945–1949 [An Ending Without Result: Sino-British Post-war Discussions over the Matter of Hong Kong, 1945–1949]* (Beijing: Social Sciences Academic Press, 2014). Other pertinent works are: Steve Tsang, *A Modern History of Hong Kong* (London and New York: I.B. Tauris &

*India, 1914–1950: A History of Imperial Diplomacy,* Lin Hsiao-ting's *Tibet and Nationalist China's Frontier: Intrigues and Ethnopolitics, 1928–49* and Chen Qianping's *Kangzhanqianhou zhi Zhong-Ying Xizang jiaoshe, 1935–1947* (Sino-British Discussions over Tibet before, during and after the Anti-Japanese War, 1935–1947).[44]

All these sources provide valuable material on Sino-British relations over the period under consideration here, but they have their limitations in the context of this study. Not only, as has been shown, is there no single study that focuses exclusively on Sino-British negotiations to secure a satisfactory bilateral post-war settlement using the three case studies chosen, but there are other shortcomings. There is, for instance, a tendency in western historiography to treat China as a passive subject of British, or American, initiatives, rather than an active actor in these events. There is also an inclination to define the Sino-British relation-

---

Co Ltd, 2004); Peter Wesley-Smith, *Unequal Treaty 1898–1997: China, Great Britain and Hong Kong's New Territories* (Hong Kong: Oxford University Press, 1980); Andrew J. Whitfield, *Hong Kong, Empire and the Anglo-American Alliance at War, 1941–45* (Basingstoke and New York: Palgrave, 2001); Kit-ching Lau Chan, "Hong Kong in Sino-British Diplomacy, 1926–45," in *Precarious Balance: Hong Kong Between China and Britain 1842–1992,* ed. Ming K. Chan (New York and London: M.E. Sharpe, 1994).

**44** Alastair Lamb, *Tibet, China & India, 1914–1950: A History of Imperial Diplomacy* (Hertingfordbury: Roxford Books, 1989); Hsiao-ting Lin, *Tibet and Nationalist China's Frontier: Intrigues and Ethnopolitics, 1928–49* (Vancouver and Toronto: University of British Columbia Press, 2006); Qianping Chen, *Kangzhanqianhou zhi Zhong-Ying Xizang jiaoshe, 1935–1947 [Sino-British Discussions over Tibet before, during and after the Anti-Japanese War, 1935–1947]* (Beijing: Sanlianshudian, 2003). Other pertinent works are: Alastair Lamb, *British India and Tibet, 1766–1910,* 2nd ed. (New York: Routledge & Kegan Paul, 1986); ibid., *The McMahon Line (I): A Study in the Relations between India, China and Tibet, 1904 to 1914,* 2 vols., vol. 1 (London: Routledge & Kegan Paul; Toronto: University of Toronto Press, 1966); ibid., *The McMahon Line (II): A Study in the Relations between India, China and Tibet, 1904 to 1914,* 2 vols., vol. 2 (London: Routledge & Kegan Paul; Toronto: University of Toronto Press, 1966); Melvyn C. Goldstein, *A History of Modern Tibet, 1913–1951: the Demise of the Lamaist State* (Berkeley, Los Angeles and London: University of California Press, 1991); Warren W. Smith Jr., *Tibetan Nation: A History of Tibetan Nationalism and Sino-Tibetan Relations* (Boulder, Colorado and Oxford: Westview Press, 1996); Amar Kaur Jasbir Singh, *Himalayan Triangle: A Historical Survey of British India's Relations with Tibet, Sikkim and Bhutan 1765–1950* (London: The British Library, 1988); Bérénice Guyot-Réchard, *Shadow States: India, China and the Himalayas, 1910–1962* (Cambridge: Cambridge University Press, 2017); Mingzhu Feng, *Zhong-Ying Xizang jiaoshe yu Chuanzang bianqing, 1774–1925 [Negotiations between China and Britain in relation to Tibet and Situations on the Tibet-Sichuan Border, 1774–1925]* (Beijing: Zhongguo Zangxue Chubanshe, 2007); Lishuang Zhu, *Minguozhengfu de Xizang zhuanshi (1912–1949) [Envoys of the Republic Governments on Their Missions to Tibet, 1912–1949]* (Hong Kong: The Chinese University Press, 2016); Gray Tuttle, *Tibetan Buddhists in the Making of Modern China* (New York: Columbia University Press, 2005).

ship merely as an interplay between China and the United Kingdom. But for most of the period Britain retained an empire and its policy towards Tibet, for example, derived primarily from the interests of British India. Further to this, Britain's China policy did not simply sway between the opinion of the Foreign Office and that of the Treasury, as is sometimes suggested.[45] British policy was driven by a mixed force of competing bureaucratic bodies, including the Board of Trade, the Colonial Office, the India Office, the Home Office, the Ministry of Labour, the Ministry of Transport, the War Office, the Admiralty, the British-Indian authorities and various British commercial groups with an interest in China, as well as key external players such as the United States. Furthermore, the signing of the Sino-British treaty in 1943 has been generally seen as a defining moment in Sino-British relations. However, while there is no denying the significance of this event, some of the "unequal" treaties were left untouched, such as those relating to Hong Kong, and in the context of discussions over the status of Tibet others took on a more complex significance. To add to this there has been a tendency to focus on the dismantling of Britain's "informal" Chinese empire in the years leading up to the period under examination. But a fuller exploration of the Sino-British discourse from 1942–1949, such as over the issues of Hong Kong and Tibet, reveals that the basis of a new bilateral relationship was far from complete by the time that the Communist regime came to power. Drawing on both British and Chinese perspectives, it is hoped that some of these shortcomings will be remedied in the study.

## Scope, Sources and Framework

As indicated above, the focus of this study is the interplay between the Chinese Nationalist government and the British government during the Pacific War and the immediate post-war period over the relinquishment of the latter's imperialist position in China and the reestablishment of a new bilateral relationship. Thus, the foreign policy of the Nationalist government, rather than the Chinese Communist regime, is the subject of this study from the Chinese perspective, given that the Nationalists were internationally recognised as the authorities responsible for China's foreign affairs in this period, including relations with Britain, and the CCP gave little thought to China's future relations with Britain before 1949. As

---

45 This view prevails in some studies of Britain's policy towards China during the period, such as Shai, *Britain and China, 1941–47*; Feng, *The British Government's China Policy, 1945–1950*; Xiang, *Recasting the Imperial Far East*.

far as the British government is concerned, the cross-party government under Conservative leadership (1940–1945) and Labour government (1945–1951) can be seen to have had virtually identical China policies.[46]

As for the timelines involved, although a movement towards revising old Sino-British treaties had already commenced in the 1920s, the Sino-British negotiations for the abolition of British extraterritoriality in China in 1942 mark a natural starting point for the discussions over post-war arrangements. Likewise, the year 1949 is regarded as the logical end for this study. From this year onwards, it was no longer the Nationalist regime, but the Communist authorities, who represented China's interest in British eyes.[47]

It should also be pointed out that while this study focuses on three major post-war Sino-British areas of contention (treaties, Hong Kong and Tibet), there were of course Sino-British discussions over a range of other issues over this period, such as military cooperation in Burma, India's independence, the distribution of US Lend-Lease goods and general war military strategy. These, however, are not included in this study because of their irrelevance to the topic of this book: the post-war settlement. It should further be noted that the three case studies chosen do not cover all the issues with regard to the post-war Sino-British settlement. For example, the non-demarcated areas of the Sino-Burmese border and the political status of Chinese immigrants in the British colonies of Southeast Asia were two other matters of dispute. However, China and Britain paid comparatively little attention to these questions after 1941 and thus relevant British and Chinese official sources are scarce. Both problems were dealt with by the Chinese Communist government in the 1950s.[48] As regards the issue of Outer Mongolia, this was a matter of dispute between China and the Soviet Union rather than Britain.

This study is based on original research in the official archives of both countries. On the Chinese side, the diplomatic documents of the Nationalist government have been consulted in the Archives of the Institute of Modern History, Aca-

---

**46** M.A. Fitzsimons, *The Foreign Policy of the British Labour Government, 1945–1951* (Notre Dame, Indiana: University of Notre Dame Press, 1953), 25–29; Joseph Frankel, *British Foreign Policy, 1945–1973* (London, New York and Toronto: Oxford University Press, 1975), 22–23; Ritchie Ovendale, "Introduction," in *The Foreign Policy of the British Labour Governments, 1945–1951*, ed. Ritchie Ovendale (Leicester: Leicester University Press, 1984), 4.
**47** Feng, *The British Government's China Policy, 1945–1950*, 170; Tang, *Britain's Encounter with Revolutionary China, 1949–54*, 193–194.
**48** The un-demarcated part of the Sino-Burmese boundary was resolved in 1961, and the Beijing government in 1955 renounced its dual-nationality policy. Those Chinese with foreign nationality then had to choose national status between China and their inhabited countries.

demia Sinica, Taipei (hereafter IMH Archives). Other Chinese documents relating to foreign and Tibetan affairs have been accessed in the collections of the Academia Historica (hereafter AH Archives) and those of the Archives of the Chinese Nationalist Party (hereafter CNP Archives). Furthermore, this study benefits from a rich array of published Chinese sources,[49] particularly those papers or diaries relating to Chiang Kai-shek,[50] Song Ziwen 宋子文 (T.V. Soong, the Chinese foreign minister, 1941–1945),[51] Wang Shijie 王世傑 (Wang Shih-chieh, the Chinese foreign minister, 1945–1948)[52] and Gu Weijun 顧維鈞 (Wellington Koo, the Chinese ambassador in London, 1941–1946).[53] On the British side, relevant documents from the Cabinet, Foreign Office, the Colonial Office and the Board of Trade have been extensively consulted in the National Archives at Kew (London), while the records of the India Office and the British-Indian government have been inspected in the Oriental and India Office Collection of the British Library.

**49** The Historical Committee of the Central Committee of the Guomindang, ed. *Zhonghuaminguo zhongyao shiliao chubian (Kangrizhanzheng shiqi): zhanshi waijiao [Preliminary Compilation of Important Historical Sources Relating to the Republic of China (the Period of the Anti-Japanese War): Wartime Diplomacy]*, 3 vols., vol. 3 (Taipei: Zhongyang Wenwu Gongyingshe, 1981) (hereafter cited as *SLCB: Wartime Diplomacy vol. 3*); The Second Historical Archives of China, ed. *Zhonghuaminguoshi dang'anziliao huibian (Nanjing Guominzhengfu 1927–1949): Waijiao [Compilation of Archives and Documents Relating to the Republic of China (the Nanjing Nationalist Government 1927–1949): Diplomacy]*, 3 vols., vol. 1–3 (Nanjing: Jiangsu Guji Chubanshe, 1997); ibid., ed. *Zhongguo Guomindang Zhongyangzhixingweiyuanhui changwuweiyuanhui huiyijilu [The Minutes of Meetings of the Standing Committee of the Central Executive Committee of the Nationalist Party]*, 44 vols., vol. 1–44 (Guilin: Guangxi Normal University Press, 2000); The Second Historical Archives of China, ed., *Zhonghuaminguoshi shiliao changbian [The Extensive Compilation of Historical Archives Relating to the History of the Republic of China]*, 70 vols., vol. 57–70 (Nanjing: Nanjing University Press, 1993); Jianlang Wang, ed. *Zhonghuaminguoshiqi waijiaowenxian huibian 1911–1949 [Compilation of Foreign Archives in the Chinese Republican Era 1911–1949]*, 10 vols., vol. 6–10 (Beijing: Zhonghua Shuju, 2015).
**50** Meihua Zhou, Sulan Gao and Jianqing Ye, eds., *Jiang Zhongzheng zongtong dang'an: shilüe gaoben [Draft Papers of Chiang Kai-shek]*, 82 vols., vol. 48–82 (Taipei: Academia Historica, 2010–2016). These were compiled and edited by Chiang's secretaries using part of his diaries and official papers.
**51** Jingping Wu and Tai-chun Kuo, eds., *Fengyunjihui: Song Ziwen yu waiguorenshi huitan jilu, 1940–1949 [T. V. Soong: Selected Minutes of Meetings with Foreign Leaders, 1940–1949 (the Chinese and English Languages)]* (Shanghai: Fudan University Press, 2010).
**52** May-li Lin, ed. *Wang Shijie riji [The Diary of Dr. Wang Shih-chieh]*, 2 vols., vol. 1 (Taipei: Institute of Modern History Academia Sinica, 2012).
**53** Wellington Koo, *Gu Weijue huiyilu [The Reminiscences of Dr. Wellington Koo]*, trans. the Institute of Modern History at Chinese Academy of Social Sciences, 13 vols., vol. 5 (Beijing: Zhonghua Shuju, 2013).

Three online databases were also consulted: the Foreign Office Files for China,[54] Hansard (the official record of all parliamentary debates)[55] and the Churchill Archive.[56] In addition, use has been made of two published archival collections: *BDFA* (British Documents on Foreign Affairs)[57] and *DBPO* (Documents on British Policy Overseas).[58] Meanwhile, American archival documents are cited mainly from *FRUS* (The Foreign Relations of the United States).[59]

Given the wide range of sources utilised, a word or two is merited on the methodology for analysis. First, only those Sino-British issues that appear in both Chinese and British official documents are covered in this study. An effort has thus been made to reconstruct events from both Chinese and British perspectives. Second, private sources and newspapers have been consulted where appropriate to counterbalance the official "top down" governmental lines which dominate the analysis. As for the overall architecture of the book, it is split into three broad parts in line with the three selected areas of analysis: treaties, Hong Kong and Tibet. Chapter I addresses Sino-British negotiations over the abolition of British extraterritoriality, while Chapter II explores the aborted efforts of Britain and China to achieve a new commercial treaty. Chapters III and IV analyse Sino-British discussion over the future of Hong Kong, while the subject of chapters V and VI are those deliberations in regard to Tibet. A conclusion then attempts to summarise the main argument in relation to the research agenda.

---

**54** http://www.archivesdirect.amdigital.co.uk.

**55** https://hansard.parliament.uk.

**56** http://www.churchillarchive.com.

**57** Anthony Best, ed. *British Documents on Foreign Affairs – Reports and Papers from the Foreign Office Confidential Print, Part III (1940–1945), Series E (Asia)*, 8 vols., vol. 5–6 (Bethesda, MD: University of America, 1997).

**58** S.R. Ashton, G. Bennett, and K. Hamilton, eds., *Documents on British Policy Overseas, Series 1, Volume 8: Britain and China 1945–1950* (London: Routledge, 2002).

**59** http://digicoll.library.wisc.edu/FRUS/Browse.html.

# Chapter I
# Negotiations over the Abolition of British Extraterritoriality and Other Related Rights in China, 1942–1943

Many books and articles have been written about the abolition of extraterritoriality in China,[1] but there is still room for further study of this aspect of the Sino-British negotiations of 1942–1943, particularly in relation to the role of the United States. As this chapter will show, it was the Chinese Nationalist government's tactic to play the American card and it was the United States, rather than Britain or China, who dominated the 1943 Sino-British treaty negotiations with the purpose of terminating British imperialism in China. More specifically, the chapter explores the processes by which the United States and Britain decided to end extraterritoriality in 1942, given that their original policy was to discuss the matter when the war ended, and how the British annulled many of their unequal treaty rights, far beyond the scope of extraterritorial rights that they had originally intended.

The chapter divides itself into several parts. After an overview of the British presence in China before the Pacific War, there is an examination of the coordinating process between the United States and Britain over the timing and the scope of the negotiations from February to September 1942. The discussion then details the Sino-British treaty negotiations in November and December 1942, during which the Nationalists managed to obtain a more extensive abolition of British treaty rights in China with the aid of the Americans.

**Acknowledgement:** This chapter is derived in part from an article of mine published in The Journal of Imperial and Commonwealth History, August 12, 2021, copyright Taylor & Francis, available online: http://www.tandfonline.com/10.1080/03086534.2021.1950326

1 See, for example, Fishel, *The End of Extraterritoriality in China*, 207–215; Kit-ching Lau Chan, "The Abrogation of British Extraterritoriality in China 1942–43: A Study of Anglo-American-Chinese Relations," *Modern Asian Studies* 11, no. 2 (1977): 257–291; Bickers, *Britain in China*, 4–5, 115–162; Li, "The Extraterritoriality Negotiations of 1943 and the New Territories," 617–650; Kayaoglu, *Legal Imperialism*, 182–189; Whewell, "British Extraterritoriality in China," 11–12, 207–215.

https://doi.org/10.1515/9783110706659-008

# 1 Britain's "Informal Empire" in China

Extraterritoriality was, in essence, the exercise of jurisdiction by one state over its own nationals within the boundaries of another. In China, British extraterritoriality was first exercised on the Chinese coast in the eighteenth century when the East India Company monopolised Sino-British trade. From the mid-nineteenth century it developed into a robust and consolidated set of legal institutions and practices.[2] Thus a legal infrastructure was established in China composed of 26 courts (1926), mainly located in treaty ports,[3] so as to try any British defendants (citizens, subjects, protégés or companies) under English, rather than Chinese, law.[4] The British Supreme Court at Shanghai was the highest British judicial body in China and considered appeals from all British courts in the country as well as in Japan (where they were abolished in 1899) and Korea (annexed by Japan in 1910).[5] As usually consuls and diplomats rendered the legal decisions, particularly in the early years, extraterritoriality was also termed consular jurisdiction.[6] Many other countries also obtained extraterritoriality in China, including the United States, Japan,[7] France, Russia, Sweden, Germany, Denmark, the Netherlands, Spain, Belgium, Italy, Austria-Hungary, Peru, Brazil, Portugal, Mexico and Switzerland.[8]

The extraterritorial rights held by the British were not reciprocated and thus deemed as an "unequal right" in Chinese eyes. To compound the matter, these rights were, in practice, frequently abused, exempting not only British nationals, but also British-protected people, and properties, vessels and companies registered as British from Chinese jurisdiction and taxation. Their weak enforcement, for example commuting or acquitting punishment of foreign culprits or criminals, caused much injustice and Chinese hatred. To complicate matters, extraterritoriality was entangled with other British privileges in China. There was the treaty-port system. After the Treaty of Nanjing (1842) five such treaty ports

---

2 Whewell, "British Extraterritoriality in China," 1.

3 Kayaoglu, *Legal Imperialism*, 4.

4 Fishel, *The End of Extraterritoriality in China*, 2.

5 Kayaoglu, *Legal Imperialism*, 6.

6 Robert Bickers, *The Scramble for China: Foreign Devils in the Qing Empire, 1832–1914* (London: Allen Lane, 2011), 5; Kayaoglu, *Legal Imperialism*, 4–6.

7 Japan and China at first reciprocally enjoyed extraterritoriality in the other's territories since 1871, but China was later deprived of the right in Japan as a result of its defeat in the First Sino-Japanese War of 1894–1895. See Dachen Cao, "Jindai Riben zai Hua lingshicaipanquan shulun [Review on Japanese Consular Jurisdiction in Modern China]," *Kangrizhanzheng yanjiu*, no. 1 (2008): 9–10.

8 Whewell, "British Extraterritoriality in China," 2.

were opened: Guangzhou, Xiamen 廈門 (Amoy), Fuzhou 福州 (Foochow), Ning-bo 寧波 (Ningpo) and Shanghai. British merchants and Chinese local authorities worked out an *ad hoc* arrangement of designating exclusive areas in major treaty ports for British residence and commerce.[9] This was later codified in the Yantai 煙臺 (Chefoo) Agreement of 1876, which provided that settlement areas should be marked out in all treaty ports. Meanwhile, as other powers followed in British footsteps, more treaty ports were designated and the British territorial presence in China expanded accordingly. This was symbolised in Britain being accorded the most-favoured-nation (MFN) status. This meant that if a Chinese treaty port was opened to a third country, Britain would have as many rights as the lat-ter in that port. In this way, the British also acquired missionary rights in the hin-terland.[10] The leaseholds in treaty ports, *de jure* under Chinese sovereignty, were governed by the British in various forms mainly based on a broader interpreta-tion of extraterritoriality.[11] Shanghai and the island of Gulangyu in Xiamen evolved into International Settlements, and were self-governed by an elected council of local residents over which the British predominated. Concessions in Tianjin, Hankou, Guangzhou, Jiujiang 九江, Zhenjiang 鎮江 and Xiamen were areas of land conceded to the British by China and then sublet. They were, for example in Tianjin, first under British consulate's direct control and later en-trusted to a municipal council organised by British locals.[12] Settlements were se-cured, such as in Wuhu 蕪湖 and Nanjing 南京 (Nanking), under which plots of land were leased to the British directly by the Chinese owners.[13] Their manage-ment was similar to other private lands held by Chinese nationals but more im-mune to Chinese authorities' intervention. Besides these settlements and conces-sions in China, Britain also retained the Crown Colony of Hong Kong and a sphere of influence in the Yangtse River (Changjiang 長江) valley, where the Brit-ish were accorded preferential rights over railway construction and mining. Over-all, these territories, either dominated or controlled by the British, thus constitut-

---

9 Jürgen Osterhammel, "Britain and China, 1842–1914," in *The Oxford History of the British Em-pire: the Nineteenth Century*, 5 vols., vol. 3, ed. Andrew Porter (Oxford: The Oxford University Press, 1999), 149–150.

10 According to Robert Neild, the total number of treaty ports was 71. Yet only a few of these ports developed into commercial and industrial centres with a strong British presence, such as Shanghai, Tianjin, Hankou 漢口 (Hankow) and Guangzhou. See Robert Nield, "Treaty Ports and Other Foreign Stations in China," *Journal of the Royal Asiatic Society Hong Kong Branch* 50 (2010): 123–139.

11 Bickers, *Britain in China*, 2; Edmund S.K. Fung, "The Chinese Nationalists and the Unequal Treaties 1924–1931," *Modern Asian Studies* 21, no. 4 (1987): 797.

12 Bickers, *Britain in China*, 137–139.

13 Nield, "Treaty Ports and Other Foreign Stations in China," 125.

ed the physical existence of Britain's "informal empire" in China, overseen by a wide judicial system backed up by extraterritoriality.

Extraterritoriality was, in many aspects, a "cornerstone" of the informal British empire in China[14] or, in Robert Bickers' words, an "instrument" of British power.[15] Under the extraterritoriality umbrella rested enormous British business interests. These were highly visible from the Chinese perspective. According to D.K. Lieu, total British capital invested in China in the late 1920s was £149.2 million: in banking (£7.5 million),[16] in insurance (£31.4 million), in manufactures (£60 million), in shipping (£15.2 million), in real estate (£10.1 million), in trade (£10.9 million), in construction (£2.9 million), in public utilities (£2.6 million) and in rubber estates and mining (£2.8 million).[17] In 1936, Britain held 35% of total foreign investments in China.[18] Shanghai and Hong Kong were the major British business hubs in China, with British investments in Shanghai ten times bigger than those in Hong Kong before 1941.[19]

There were further British privileges in China. Through the Treaty of the Bogue (1843) and the Treaty of Tianjin (1858), British warships and civilian vessels acquired the right to sail in Chinese coastal and inner waters (mainly the Yangtse River).[20] Thereafter, shipping gradually developed into one of Britain's principal commercial activities in China. British shipping companies came to

---

**14** Jürgen Osterhammel, "China," in *The Oxford History of the British Empire: the Twentieth Century*, 5 vols., vol. 4, ed. Judith M. Brown and William Roger Louis (Oxford: Oxford University Press, 1999), 653.

**15** Bickers, "Legal Fiction," 53.

**16** British financial interests in China were mainly entrusted to two British banks: the Hong Kong and Shanghai Banking Corporation (HSBC) and the Chartered Bank of India, Australia and China (CBIAC). See Niv Horesh, *Shanghai's Bund and Beyond: British Banks, Banknote Issuance, and Monetary Policy in China, 1842–1937* (New Haven and London: Yale University Press, 2009), 11–15.

**17** D.K. Lieu, *Foreign Investments in China* (Nanjing: Chinese Government Bureau of Statistics, 1929), table 39, 115–116, cited in Horesh, *Shanghai's Bund and Beyond*, 42–43. C.F. Remer has estimated total British direct investments in China proper in 1931 at £197.9 million. See C.F Remer, *Foreign Investments in China* (New York: H. Fertig, 1933), table 14, 397, quoted in Horesh, *Shanghai's Bund and Beyond*, 43.

**18** Bickers, *Britain in China*, 10–13.

**19** William Roger Louis, "Introduction," in *The Oxford History of the British Empire: the Twentieth Century*, 5 vols., vol. 4, ed. Judith M. Brown and William Roger Louis (Oxford: Oxford University Press, 1999), 41.

**20** Yumin Li, "Wanqing Zhong-wai Bupingdengtiaoyue de falü guanxi [The Legal Connections among the Unequal Treaties of the Late Qing Era]," in *Jindai Zhong-wai guanxishi yanjiu [Study on the History of Modern Sino-Foreign Relations]*, vol. 7, ed. Junyi Zhang and Hongmin Chen (Beijing: Social Sciences Academic Press, 2017), 8.

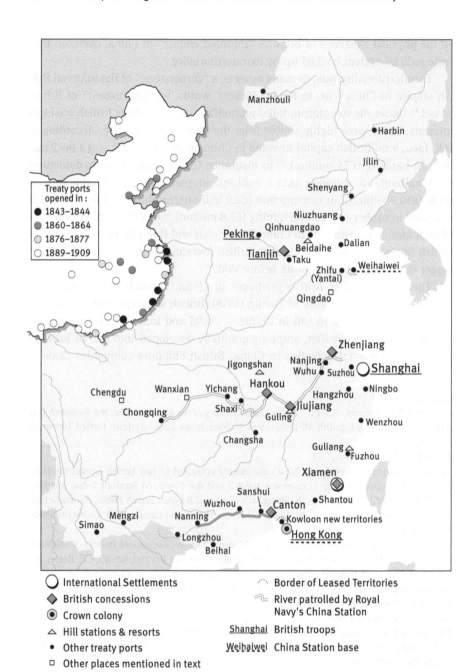

**Map 1:** Britain in China, January 1, 1927.
Source: Bickers, "Britain in China, 1 January 1927," in Bickers, *Britain in China*, 11.

carry about a third of China's coastal trade and rather more trade on the Yangtse River. Two companies, Jardine, Matheson & Co. and Butterfield & Swire, special-ised in this area.[21] The Chinese Maritime Customs Service also came to be domi-nated by the British. As a result of the Treaty of Nanjing (1842), China was de-prived of "tariff autonomy" and had to negotiate with Britain over tariffs and similar charges. Later, the Shanghai Customs House, a temporary arrangement established by the British during the Taiping Rebellion (1851–1864), was accept-ed by the Qing government as a model to be applied throughout China. Hence, a Chinese maritime customs bureaucracy was formed in 1861–64, headed by the British and mainly staffed by foreigners.[22] In 1898, the dominant role of the Brit-ish was further confirmed by a commitment from the Qing court that a Briton would hold the position of Inspector General as long as British trade with China was in the preponderance.[23] The power of the Service expanded rapidly. It became "a debt-collection agency for foreign bondholders" since a large amount of foreign loans was hypothecated on customs revenue to pay for the First Sino-Japanese War (1894–1895) and the Boxer War (1900–1901) indemni-ties.[24] The 1911 revolution marked a further turning point for the Service, trans-forming it from a branch of the Chinese government into an *imperium in imperio*. During the revolution, the Service alienated itself from the Qing government and directly controlled customs revenue collection in order to secure the payment of China's loans and indemnities. The embryonic republican government was too weak to prevent this. Conversely, it heavily depended on the Service's financial support to overpower numerous provincial rivals and the Inspector General con-sequently became China's "Finance Supremo".[25] Under the Nationalists, the Chi-nese government regained tariff autonomy. Yet the British domination of the Service generally remained intact although it was nominally re-affiliated to the Chinese Finance Ministry.[26] To complete the picture of privileges, the British op-erated a legation quarter in Beijing 北京 (Peking); oversaw the international management of the Hai River 海河 (Peiho) in Tianjin and the Huangpu River 黄浦江 (the River Whampoo or Whangpoo) in Shanghai; had the right of the em-

---

**21** TNA, FO 371/69559, telegram from D.F. Allen (UK shipping Representative for the Far East) to F.V. Cross (British Embassy, Washington), January 7, 1948.
**22** The British remained in the position of Inspector General until 1943; van de Ven, *Breaking with the Past*, 26–39.
**23** Bickers, "Revisiting the Chinese Maritime Customs Service, 1854–1950," 221–222.
**24** van de Ven, *Breaking with the Past*, 133–134.
**25** Ibid., 172–173.
**26** Chen, *Zhongguo jindai haiguan shi*, 634–638, 673–675.

ployment of foreign pilots in Chinese coastal and inner waters, railway and mining rights; and enjoyed the freedom of conducting missionary work in China.[27]

China's struggle to abolish the unequal treaty rights started in the late nineteenth century. In 1902–1903, Britain, the United States and Japan agreed in principle to extraterritoriality being removed provided that an adequate judicial and penal system, in accordance with that of western nations, was established to protect the legal and property rights of their citizens.[28] Yet, a Chinese request for the abolition of extraterritoriality was refused at the Paris Peace Conference (1919) and the Washington Conference (1921–1922) on the grounds that the judicial conditions failed to meet the requirements of Britain and the other powers,[29] despite the fact that China had contributed much to the Allied powers' victory in the First World War.[30] In 1926, a commission on extraterritoriality, formed in the wake of the Washington Conference, was appointed to look into the Chinese judicial system. However, it was deemed that extraterritoriality should remain in view of China's fragmented and under-funded legal bureaucracy and its shortage of competent, trained, independent judges.[31] After it came to power in 1926–1927, the Guomindang regime, drawing on the public mood of nationalism and anti-imperialism, continued to wage a campaign to revise the existing treaties with foreign countries, including those relating to extraterritoriality. In 1931 a Sino-British agreement over the abolition of extraterritorial rights, based on the condition that the International Settlement (Shanghai) and British Concession of Tianjin would be excluded for ten years, almost materialised but failed due to a split in the Guomindang, Japan's invasion of Northeast China in Septem-

---

**27** Li, "The Extraterritoriality Negotiations of 1943 and the New Territories," 618–619.

**28** Fishel, *The End of Extraterritoriality in China*, 27.

**29** Kayaoglu, *Legal Imperialism*, 161.

**30** Between 1915 and 1921 at least 300,000 Chinese labours joined the war effort of the Allied powers. Nearly 100,000 Chinese were recruited into the British Chinese Labour Corps of whom over 2,000 died. France recruited about 36,700 Chinese labours while Russia utilised some 160,000. See Brian C. Fawcett, "The Chinese Labour Corps in France 1917–1921," *Journal of the Hong Kong Branch of the Royal Asiatic Society* 40 (2000): 50; Christian Koller, "The Recruitment of Colonial Troops in Africa and Asia and Their Deployment in Europe during the First World War," *Immigrants & Minorities* 26, no. 1–2 (2008): 113; Olga Alexeeva, "Chinese Migration in the Russian Far East: A Historical and Sociodemographic Analysis," *China Perspectives*, no. 3 (2008): 23–24.

**31** Kayaoglu, *Legal Imperialism*, 171–173.

ber 1931 and the organised resistance from British settlers mainly in Shanghai.[32] It would take the coming of the Pacific War to free the bottleneck.

## 2 The Pacific War and the Revival of the Abolition of Extraterritoriality

In December 1941, the United States, as well as American and British colonies in Southeast Asia, were attacked by the Japanese. This eventually led to their military alliance with China, which had already been engaged in a total war against Japan alone for four years. However, the United States and Britain were not prepared to immediately mount a counterattack against Japan. The war industries of the United States needed time to be fully mobilised and, compared to Germany and Italy, Japan was a relatively less immediate threat to Britain. Thus, the Alliance's strategy was to defeat Nazi Germany and Fascist Italy first and then turn on Japan. Because of this, China's continued war role was important since it tied up large numbers of Japanese military forces. For his part, Chiang Kai-shek was encouraged by the entry of the United States and Britain into the Pacific War. His regime had been on the edge of bankruptcy in 1941 and was eager to receive financial and logistical support from the United States and Britain. Nevertheless, in tandem with seeking loans and Lend-Lease goods from Washington and London, the Chinese tentatively drew attention to the abolition of extraterritoriality.

Anglo-American policy had been to avoid discussion of the matter until the end of Sino-Japanese hostilities. Winston Churchill, the British Prime Minister, reiterated this policy in the House of Commons on July 18, 1940:

> We are ready to negotiate with the Chinese Government, after the conclusion of peace, the abolition of extraterritorial rights, the rendition of concessions and the revision of treaties on the basis of reciprocity and equality.[33]

---

32 Fung, "The Chinese Nationalists and the Unequal Treaties 1924–1931," 815–817; Fishel, *The End of Extraterritoriality in China*, 183, 186–187; Chan, "The Abrogation of British Extraterritoriality," 258–259; Bickers, *Britain in China*, 116, 152–153; Kayaoglu, *Legal Imperialism*, 181–182.
33 *House of Commons Debates*, vol. 363, July 18, 1940, col. 400,
   http://hansard.millbanksystems.com/commons/1940/jul/18/transit-of-war-material-to-chi na#S5CV0363P0_19400718_HOC_238 (accessed November 11, 2017). Before the statement, messages of the same meaning had been sent to the Chinese government on January 14, 1939 and June 11, 1941. See Fishel, *The End of Extraterritoriality in China*, 208; Li, "The Extraterritoriality Negotiations of 1943 and the New Territories," 619.

The Americans held the same view. In April 1941, Guo Taiqi 郭泰祺 (Quo Tai-chi), the Chinese ambassador to London from 1932–1941, was sent by Chiang Kai-shek to the United States to discuss the question of extraterritoriality *en route* to taking up his new post of Foreign Minister in Chongqing.[34] Cordell Hull, the American Secretary of State, confirmed to him that the negotiations would recommence when peace would have been restored to China.[35]

Chiang Kai-shek was not, however, deterred from keeping the issue on the Allied agenda. His most famous messenger was Madame Chiang Kai-shek, Song Meiling 宋美齡. In a series of articles and broadcasts directed at American audiences, she brought the matter of extraterritoriality and other foreign privileges in China to the attention of the American public. In a seminal article in *The New York Times* of April 19, 1942, she wrote:

> In the first [stage] the weapon of the West toward China was always force. By the pointing of a gun at her the West made her suffer humiliation after humiliation. All her port cities were opened, in an actual as well as a metaphorical sense, at the point of the bayonet. [...]
>
> In the meantime, the West established self-governing cities in China on the West's own model in violation of China's sovereign rights, but as a face-saving gesture shrouded them under a thin veil of foreign settlements and concessions. The West also instituted the vicious legal device known as extraterritoriality, which removed foreigners from the jurisdiction of Chinese courts.
>
> Nor did the West keep its hands off our material resources. The richest of our mines passed under foreign control. Foreigners administered our customs, salt revenue, railways, in fact, took over the management of virtually all public utilities, while even the control of foreign exchange was vested in them. With the exception of the Christian church, the policy of the West, on the whole, seemed to be to get as much as possible from us by force and to give nothing in return that it could withhold.[36]

The voice of China's first lady, who had been educated in the United States and had graduated from Wellesley College, was influential. According to the comment by one American official in May 1942, "there were a considerable number

---

34 *Waijiaobu Guo Taiqi zi Huashengdun zhi Waijiaobu gaoyi yu Mei waizhang guanyu feichu bu-pingdetiaoyueshi zhi huanwen neirong dazhi dian* [The telegram from Guo Taiqi to the Chinese foreign Ministry about the content of an exchange of notes with America on the abolition of extraterritoriality], May 25, 1941, The Historical Committee of the Central Committee of the Guomindang, ed., *SLCB: Wartime Diplomacy vol. 3*, 708.

35 *Meiguo waizhang fuhan* [The Reply from the American Secretary of the State], May 25, 1941, ibid., 709. Similar to the Sino-British one, the Sino-American negotiations over the abolition of extraterritoriality had suspended in 1931.

36 Meiling Song, "First Lady of the East Speaks to the West," *The New York Times*, April 19, 1942, https://search-proquest-com.ezproxy.is.ed.ac.uk/docview/106203461?accountid=10673.

of American writers and others who were exhibiting active interests in the matter [of extraterritoriality]" and the United States government "should not be surprised if pressure upon the Department to take action [to abolish extraterritoriality] increased."[37] Nevertheless, the Nationalists never officially raised the matter during this period, probably since they had no wish to deviate from the main task of enlisting the United States' military support.

## 2.1 Anglo-American Coordination over Early Abolition

As the war in the Pacific intensified, the American government began to push for an early abolition of extraterritoriality in China. This attitude was partly driven by the importance of China's role in confronting the Japanese in the absence of a strong American military presence in the region, but also by anti-imperialist sentiments among American officials. In a series of memoranda drawn up by the Division of Far Eastern Affairs at the Department of State, American diplomats weighed up the conditions for an early abolition of extraterritoriality. Maxwell Hamilton, the Chief of the Division of Far Eastern Affairs, admitted that the time was not ideal, at least not until the question of the future status of India had been resolved by the British government, but he argued that the pressures for early abolition were of "a more substantial and enduring character than those contras."[38] On his advice, the Division of Commercial Treaties and Agreements and the Division of Far Eastern Affairs set up a small inter-departmental committee to begin preparatory work, in strict confidence, on the draft of a suitable treaty. But some US officials were hesitant. Stanley Hornbeck, an adviser on Political Relations in the Department of State, deemed there was no good reason for the Americans to play this card at the current time since "the subject of our extraterritorial rights in China is not at this moment vividly in the minds either of the Chinese or of our own people."[39]

Concurrently, Anthony Eden, the British Foreign Secretary, was taking stock of the issues involved in early abolition. The aims of the British in this regard were the "encouragement of the Chinese as a preparation for work together in

---

37 Memorandum of Conversation, by the Assistant Chief of the Division of Far Eastern Affairs (Atcheson), May 8, 1942. US Department of State, ed. *Foreign Relations of the United States Diplomatic Papers, 1942, China* (Washington, D.C.: Government Printing Office, 1960), 279–280 (hereafter cited as *FRUS: 1942 China*).
38 Memorandum by the Chief of the Division of Far Eastern Affairs (Hamilton), March 27, 1942. Ibid., 271–274.
39 Memorandum by the Adviser on Political Relations (Hornbeck), April 9, 1942. Ibid., 274–275.

the post-war era and as a counterpoise to a possible agreement with Russia."[40] Eden, however, questioned whether the Chinese would regard abolition as a generous gesture, and react accordingly, or take it as an intimation of British weakness. In order to make things clearer, he requested the opinion of Horace Seymour, the British Ambassador to Chongqing. Seymour encouraged the government in Westminster to pursue the matter but stressed that the Chinese would be unlikely to accept any new treaty provisions other than those that were usually included in treaties between equal nations. He also drew attention to the erosion of British interests in China. Wartime regulations, for example those over banks, were being enforced by Chongqing without consultation with the British government. The British, moreover, had no good reason to protest against these new Chinese powers. There was little British trade and commerce now in China, in either the free or Japanese occupied areas. And, in light of the strong Chinese patriotism unleashed by the Japanese war and Britain's poor military performance in Southeast Asia, any British opposition to early abolition would only stimulate fresh anti-British sentiment among the Chinese population.[41]

Nonetheless, London remained stubborn on the issue. Eden came to the conclusion that the present moment was not suitable for taking the initiative over extraterritoriality. Instead, the British would reconsider it when the military tide turned against Japan.[42] They would respond sympathetically if the Chinese government decided to raise the issue, but, as Seymour suggested, the Chinese government might deliberately avoid it.[43] In these circumstances, the only real concern from the British perspective was independent American action. In order to forestall this possibility, Whitehall felt it necessary to secure a promise of prior consultation if the United States government itself desired to act.

Events were now moving apace. Hamilton was assigned to revisit the matter. In a conversation of May 1942 with H. Ashley Clarke, the head of the Far Eastern Department of the Foreign Office (1942–1944), during the latter's short visit to Washington, he was eager to know whether the British had given any thought of issuing a declaration, without any qualification, that upon the restoration of

---

**40** F 2031/74/10, "Mr Eden to Sir A. Clark Kerr (Chongqing)", March 28, 1942, Anthony Best, ed. *British Documents on Foreign Affairs – Reports and Papers from the Foreign Office Confidential Print, Part III (1940–1945), Series E (Asia)*, 8 vols., vol. 5 (Bethesda, MD: University Publications of America, 1997), no. 29, 37 (hereafter *BDFA, Part III, Series E, vol. 5*).
**41** TNA, FO 371/31679, telegram from H. Seymour to Robert Anthony Eden, April 1, 1942.
**42** F 2757/828/10, "Mr Eden to Viscount Halifax (Washington)", April 20, 1942, Best, ed., *BDFA, Part III, Series E, vol. 5*, no. 4, 111.
**43** Li, "The Extraterritoriality Negotiations of 1943 and the New Territories," 624.

peace Britain would relinquish its extraterritorial rights in China; or, as an alternative, concluding a treaty with China that would simply contain a future commitment. Clarke, however, deemed that neither measure was advisable, especially in view of the many military reverses that had been experienced in East Asia at this time.[44] In the meantime, the American position was hardening. In a telegram from Cordell Hull to London in July, the American stance was now to relinquish their extraterritorial rights "at the earliest moment", but not "at the most advantageous time" from the British point of view.[45]

The US government moved quickly towards a pro-abolition stance. Hull considered that "the present is probably as good a time as any, especially as the initiative still lies with us", and hoped to accomplish three principal objectives through abolition:

(a) some psychological and political benefit to the cause of the United Nations which would be of concrete assistance to China and thus tend to strengthen the determination of that country in its war effort;
(b) the wiping out once and for all of an existing anomaly in our relations with China;
(c) the achievement of agreement in principle to regularise in China the usual rights normally accruing to American and British nationals in friendly foreign countries.[46]

With regard to the process of abolition, a treaty, rather than just a declaration to abolish rights, was more preferable to Washington. Through a formal treaty, Hull hoped, not only would the terms of the termination of extraterritorial rights and practices in China be properly laid out, but also an outline of the general framework of a relationship in the post-war world. Furthermore, in view of the tense military position at that particular time, the treaty must be a brief one. A more comprehensive treaty would probably be not completed for several months and thus be hard to be kept confidential for such a long time. "The enemy governments might be expected to seize every available opportunity to criticize to their own advantage any delay in negotiations or any apparent differing of views between the parties to the negotiations."[47]

---

**44** Memorandum of Conversation, by the Chief of the Division of Far Eastern Affairs (Hamilton), May 22, 1942, US Department of State, ed., *FRUS: 1942 China*, 280.
**45** The Secretary of State to the Ambassador in the United Kingdom (Winant), July 11, 1942, ibid., 281–282.
**46** The Secretary of State to the Ambassador in the United Kingdom (Winant), September 5, 1942, ibid., 287–288.
**47** The Secretary of State to the Ambassador in the United Kingdom (Winant), August 27, 1942, ibid., 282–286.

Washington consulted London. John G. Winant, the American ambassador, tried to persuade Clarke of the efficacy of the Americans' intentions. Yet the latter's unwillingness was still evident. Clarke remarked that a short time previously both governments had agreed that the time was not opportune to raise the question and he was curious about exactly what new elements had arisen to cause the Americans to change their view. Furthermore, he did not think there was much likelihood that the Chinese themselves would raise the issue now, asserting that Chongqing had always taken the position that extraterritorial rights belonged to the past. Without other effective ways to persuade Clarke, Winant could only ascribe the change of heart in Washington to recent military progress in the Solomon Islands, in addition to the pressure of American public opinion.[48]

Indeed, not every American official was in favour of abolition at this time. Clarence E. Gauss, the American Ambassador to Chongqing, was one of the dissenters. In his view, "an over-generous policy at this time of first surrendering extraterritorial and related rights and then expecting later fair and just treatment in general and trade relations would be fatal." Hence, he recommended a treaty with adequate protection of existing American interests and future business opportunities in China.[49] Nevertheless, the final decision was not decided at an ambassadorial level. Washington still wished for a quick treaty without too many strings. In these circumstances, Eden decided to bring the British into the negotiations.

## 2.2 Anglo-American Concurrence over the 1943 Treaty

Anglo-American discussions over the contents of the treaty were held in strict confidence and no evidence exists to show that the Chinese government was aware of the negotiations. Under these conditions, the US Department of State and the UK Foreign Office cooperated on the timing and the scope of a treaty and were finally in accord by the autumn of 1942.

According to Winant, the British desired to "limit action at present to the conclusion of a brief treaty dealing only with the abrogation of extraterritorial rights and the related questions". They did not expect abolition to extend to

---

**48** The Ambassador in the United Kingdom (Winant) to the Secretary of State, September 1, 1942, ibid., 286. The military progress referred to here was related to the Guadalcanal campaign. In August 1942, the Allied forces, predominantly United States Marines, reoccupied the principal island of the Solomon Island, Guadalcanal.
**49** The Ambassador in China (Gauss) to the Secretary of State, September 8, 1942, ibid., 288–290.

other British rights in China. If the Chinese pressed for a more comprehensive agreement, it was suggested that "the two Governments should insist on the larger questions being postponed until the end of the war."[50] Eden also argued, as had his predecessors in 1931, for some special arrangement for Shanghai "to enable the development of the port to continue with the cooperation of foreign commercial interests."[51] It was suggested that this possibility had actually been raised by Wang Pengsheng 王芃生, Chiang Kai-shek's leading advisor on Japanese affairs and head of the Institute for International Relations at the Military Affairs Commission (MAC) of the Nationalist government.[52] Yet the Chinese were unlikely to accept such a condition since most of Britain's "informal empire" in China had already been destroyed by the Japanese and, while China urgently needed British aid, the British required Chinese military assistance to defend Burma and India. It also seemed improbable that Chiang Kai-shek would countenance any continuance of international control over the administration of Shanghai, with the exception of few foreign advisers. Given these considerations, and in view of the fact that Washington was inclined to "the complete wiping out of all rights of a special character" connected with extraterritoriality,[53] the British government agreed to "avoid any specific provision in the proposed treaties which would tie its hands with respect to Shanghai."[54]

In October 1942 the Americans and the British agreed a draft treaty. It focused on the abolition of extraterritoriality, but also included related treaty rights such as the return of the International Settlements in Shanghai and Gulangyu (without influencing the existing ownership of properties in China possessed by foreign nationals and corporations) and other minor treaty rights, such as the termination and return of the Diplomatic Quarter at Beijing. In addition, London would forsake its concessions in Guangzhou and Tianjin.[55]

The British suggested opening negotiations with Chinese at this stage. The best approach, they believed, would be a joint declaration issued by the two gov-

---

50 The Ambassador in the United Kingdom (Winant) to the Secretary of State, September 8, 1942, ibid., 291–292.
51 Ibid.
52 The Ambassador in the United Kingdom (Winant) to the Secretary of State, September 15, 1942, US Department of State, ed., *FRUS: 1942 China*, 293–295.
53 The Secretary of State to the Ambassador in the United Kingdom (Winant), September 12, 1942, ibid., 292–293.
54 The Secretary of State to the Ambassador in China (Gauss), September 18, 1942, ibid., 295–296.
55 The Ambassador in the United Kingdom (Winant) to the Secretary of State, October 6, 1942, ibid., 302–303.

ernments on China's National Day (October 10).[56] Such a declaration would hopefully enhance its value in Chinese eyes. Hull saw the advantages of this suggestion but refused to agree on the grounds that "joint announcements [drafted by the British] might carry an implication of pressure upon the Chinese Government."[57] Consequently, the Americans judged that individual announcements by the two governments would be preferable and the British accepted this course of action.

## 3 Sino-British Negotiations and the Abolition of Other British Privileges in China

An interesting distinction concerning the naming of the 1943 Treaty has been consistently neglected in scholarly works. In China, it has frequently been termed the new "equal" Sino-British treaty (*Zhong-Ying pingdeng xinyue* 中英平等新約), while western historians tend to focus on extraterritorial rights and the regulation of related matters.[58] In fact, neither of these terms is accurate. The Chinese negotiators did not deem the treaty equal since the British were stubborn over the issue of the New Territories of Hong Kong and entangled the discussions with too many safeguarding clauses. They accordingly extracted the word "equality" from their original draft and no such wording was allowed to appear in the formal version. London was also disappointed with the outcome. Its initial effort to limit the negotiations to the matter of extraterritoriality was largely undermined by China's desire for full abolition and Washington's supportive attitude. As a result, Whitehall was compelled to annul many of its rights in China. The distinctive terminologies only reflected the different initial aims of the Chinese and British for the negotiations and neither Chongqing nor London was satisfied with the outcome.

This unsatisfactory outcome could be traced to the very beginnings of the negotiations. It was agreed that the British and the Americans would communicate with the Chinese *Chargé d'affaires* in London and the Chinese Ambassador in Washington respectively on October 9 and then declare their separate notifications to the press on October 10. However, in the event, the British declaration

---

56 Ibid.
57 The Acting Secretary of State to the Ambassador in the United Kingdom (Winant), October 7, 1942, US Department of State, ed., *FRUS: 1942 China*, 304–305.
58 The English title of the treaty is "Treaty for the Relinquishment of Extra-Territorial Rights in China and the Regulation of Related Matters". Robert Bickers terms it in one of his papers "the Sino-British Friendship Treaty". See Bickers, "Legal Fiction," 53.

arrived late at Chongqing for unspecified reasons. This delay planted suspicion in the Chinese leader's mind, even before the negotiations had started.[59]

## 3.1 Chinese Aspirations for Full Abolition and the United States' Endorsement

On October 9, one day before China's National Day, the news of the intention of the United States and the United Kingdom to restart extraterritoriality negotiations arrived in Chongqing. Chiang was delighted but immediately instructed Song Ziwen, the Chinese Foreign Minister then in Washington, to persuade the Americans into accepting a broader abolition of privileges such as concessions and settlements, the rights of foreign troops stationed in China, inland navigation and the customs service. In his mind, only if these privileges were included would abolition be sincere and truthful.[60] Yet, sounding a note of caution, Chiang debarred Song from any official discussion with the Americans before their draft treaty was released. In a telegram of October 12, he suggested that the latter conduct any communication "indirectly".[61] In tandem with this, he tentatively publicised his aspiration for a wider agreement through a governmental gazette in which the Americans were encouraged to "unilaterally and voluntarily abolish all unequal clauses in the Sino-American treaties."[62] Chiang did not deliver such a message to the British. His tactic was to wait for an American general abolition first and then to push London to follow suit. After receiving the American treaty draft, the Chinese government pressed for the ending of treaty rights relating to the "treaty-port system, special courts at Shanghai, the carrying on of coastal trade and inland navigation by foreign nationals, the employment of for-

---

59 Koo, *Gu Weijue huiyilu vol. 5*, 100.
60 Telegram from Chiang Kai-shek to Song Ziwen, October 10, 1942, Meihua Zhou, ed., *Jiang Zhongzheng zongtong dang'an: shilüe gaoben [Draft Papers of Chiang Kai-shek]*, 82 vols., vol. 51 (Taipei: Academia Historica, 2011), 386–387 (hereafter cited as *SLGB vol. 51*).
61 *Jiangweiyuanzhang zi Chongqing zhi Waijiaoabuzhang Song Ziwen zhu guanyu feichu zhiwaifaquan shi ying jingdai Meizhengfu tichu jianduan zhi caoyue hou wofang zaixing biaoshi yijian dian* [Telegraph from Chiang Kai-shek to T.V. Soong that the Chinese government should wait for an American draft and then express its opinion], October 12, 1942, The Historical Committee of the Central Committee of the Guomindang, ed., *SLCB: Wartime Diplomacy vol. 3*, 714.
62 *Junshiweiyuanhui Shicongshi di'erchu zhuren Chen Bulei zuanni xiwang Meiguo shuaixian zidong biaoshi fangqi duihua bupingdengtiaoyue xinwengao* [Instruction to Chen Bulei about drafting a news report that the Chinese government wished for an American initiative to abolish its unequal treaty rights in China], October 5, 1942. Ibid., 710.; Zhou, ed., *SLGB vol. 51*, 349–352.

eign pilots, and the entering and anchoring of foreign warships in Chinese ports without previous consent."[63]

Washington was sympathetic to a broader abolition and agreed to deal with these additional issues in an exchange of notes affiliated with the treaty.[64] While its cultural and missionary presence in China was extensive, the US had fewer commercial interests in comparison to Britain. Its only two settlements in Shanghai and Gulangyu were shared with other powers and its economic concerns concentrated in the import-export business. Standard-Vacuum Oil was the largest American firm in China and its sales network in China formed the largest single item of American direct investment. Little or no American capital was invested in Chinese mining, shipping or manufacturing. In 1929–1930 two of the most profitable public utility companies in Shanghai, the Yangshupu 楊樹浦 Power Plant and the Shanghai Telephone Company, were transferred from British to American ownership. Yet American business interests in China were still "underdeveloped and dispersed".[65]

The American attitude put the British on the back foot. In this new epoch, the traditional gunboat policy or other punitive measures were no longer viable. Instead, Whitehall resorted, ironically, to the principles of equality and reciprocity in order to safeguard its British treaty rights in China.

### 3.2 "The Principles of Equality and Reciprocity"

The equal and reciprocal principles were also the rationales behind Chinese aspirations for terminating the existing treaty system. The Chinese government at first aspired to insert a sentence of this effect into the opening paragraph of the treaty. Interestingly, this proposal encountered objections from the United States but not Britain.[66] Washington could accept the equality doctrine, but not that of

---

63 *Waijiaobu Song Ziwen zi Chongqing cheng Jiang weiyuanzhang baogao Waijiaobu dui Zhong-Meiguanxi tiaoyue cao'an yijian ji gaibu zhengli zhi Zhong-Meiguanxi tiaoyue xiuzheng'an qiancheng* [Comments of the Chinese Foreign Ministry on the American treaty draft], The Historical Committee of the Central Committee of the Guomindang, ed., *SLCB: Wartime Diplomacy vol. 3,* 722–728.

64 *Zhumeidashi Wei Daoming zi Huangshengdun zhi Waijiaobu baogao yu Mei waijiaobu jiu Zhong-Meitiaoyue zengshan gedian zuo feizhengshi taolun zhi jingguo dian* [The telegram from Wei Daoming to the Chinese Foreign Ministry about American views on the Chinese-revised Sino-American treaty draft], ibid., 728–729.

65 Osterhammel, "China," 643–644.

66 TNA, CO 129/588/24, "Cypher telegram from Secretary of State for India to Government of India, External Affairs Department", November 25, 1942.

reciprocity. At that time, the Chinese Exclusion Act was still in existence in the United States, and it would have been hard for the American government to induce Congress to abolish such legislation. Furthermore, at the state level, both California and Oregon forbade the Chinese to possess property.[67]

It was also not straightforward for Britain to reciprocate in this way, although in this case the reason was largely due to the complex arrangements of the British Empire and its numerous dominions and colonies. For example, in Kenya, there were restrictions on non-European ownership of lands in the Kenyan Highlands. If the British assented to opening up such a land market to the Chinese, it would arouse substantive protests from the Indian or other Asian and African colonial authorities.[68] Furthermore, both India and Burma were anxious about an influx of Chinese investors looking to purchase property, which in Burma had already occurred during the Sino-Japanese War before 1941.[69]

Nonetheless, London was still hopeful that by offering China equal and reciprocal rights across the British Empire – an equivalent concession that the United States was not prepared to offer – Chongqing would agree to the continuation of Britain's favourable status within China. Yet notwithstanding their frequent use of the terms "equality" and "reciprocity" to criticise the existing unequal treaty system, the Nationalists now saw these principles as a hinderance to a full abolition of British privileges in China. At this point, their attitude was little different from that of their Communist rivals. Mao Zedong, the founder of the People's Republic of China (PRC), indicated vividly in the 1950s that the policy of his regime towards foreign imperialism was "cleaning the house before entertaining guests."[70] In the end, the Nationalists dropped their original propos-

---

**67** In 1943, after the signing of the 1943 Sino-American Treaty, Congress repealed the Exclusion Act.

**68** TNA, CO 129/588/24, letter from Ashley Clarke (FO) to G.F. Seel (CO), December 3, 1942.

**69** TNA, CO 129/588/24, telegram from W. Johnston (BO) to Ashley Clarke (FO), November 26, 1942.

**70** The policy was reiterated by Zhou En'lai, the Chinese Prime Minister, and Shi Zhe, Mao's translator of the Russian language. See En'lai Zhou, "Our Foreign Policy and Mission (April 30, 1952)," in *Zhou En'lai Xuanji [Selected Collection of Zhou En'lai's Speeches and Articles]*, 2 vols., vol. 2, ed. Zhonggong Zhongyang Wenxian Yanjiushi Bianji Weiyuanhui (Beijing: Renmin Chubanshe, 1984), 85–89; Zhe Shi, *Zai lishi juren shenbian [Standing Beside Giant Man in History]* (Beijing: Zhongyangwenxian Chubanshe, 1991), 379–380.

al of putting some wording to the effect of equality and reciprocity into the trea-ty.[71]

## 3.3 Negotiations over British Safeguarding Clauses

British objections during the Sino-British treaty negotiations were not primarily concerned with the treaty-port system and British dominance of the Chinese Maritime Customs Service. The treaty-port system was no longer indispensable since the Chinese government was willing to open all Chinese territories to British nationals for residence and commerce[72] and grant American and British vessels both national treatment and MFN treatment in terms of overseas merchant shipping,[73] except for reserving the right to close any ocean port on the grounds of national security.[74] It was also the case that the position of the Chinese Maritime Customs Service had significantly declined since the start of the Sino-Japanese War (1937–1945). The major part of the Service was now controlled by the Japanese in occupied China, from which over 80% of customs revenue was derived.[75] This left the Service in Chongqing without any significant tax income, especially after the Inter-port Duty (1931–1941) and its replacement, the Wartime Consumption Tax (1942–1944), were abolished.[76] The Chinese government thus proposed merely to terminate the British right to claim the appointment of a British subject as Inspector General of the service while the volume of British trade exceeded that of other nations. Thus, the major Sino-British debates concentrated on the introduction of British safeguarding clauses in light of the decision to abolish extraterritoriality and its related rights.

---

**71** TNA, CO 129/588/24, telegram from Sir H. Seymour to Foreign Office, December 15, 1942. The use of "equality" was retained in the Sino-American treaty. See Appendix II, Fishel, *The End of Extraterritoriality in China*, 233–240.

**72** TNA, CO 129/588/24, telegram from Foreign Office to Chongqing, November 26, 1942.

**73** See Clause 1(a) in Annex of Exchange of Notes and Clause 1 in Agreed Minute, available from UK Treaties Online, "Treaty between His Majesty in respect of the United Kingdom and India and His Excellency the President of the National Government of the Republic of China for the Relinquishment of Extra-Territorial Rights in China and the Regulation of Related Matters (with Exchange of Notes and Agreed Minute)," 1943, https://treaties.fco.gov.uk/awweb/pdfopener?md=1&did=64833 (accessed March 28, 2019).

**74** TNA, CO 129/588/24, telegram from Sir H. Seymour to Foreign Office, December 25, 1942.

**75** Boecking, *No Great Wall*, 203.

**76** Ibid., 205–212; van de Ven, *Breaking with the Past*, 281–286.

### 3.3.1 Personal Status and Dual Nationality

The first safeguard the British endeavoured to put into the treaty was a personal status clause. It was proposed that "in matters of personal status [the] law of territory from which [the] party concerned originates will be applied by the courts of one of the High Contracting Parties as regards national designations under the jurisdiction of the other High Contracting Party."[77] Simply put, the Chinese courts would have jurisdiction over British nationals in China but had to apply English laws to them. However, London could not reciprocate the same treatment to the Chinese in Britain since it operated a law of domicile.[78] Unsurprisingly, Chongqing refused this proposal and Britain had no choice but to set the matter aside for later discussion.[79]

Another matter that concerned London was dual nationality, a status possessed by a large number of Chinese immigrants living in British colonies of Southeast Asia. According to the Nationalist government's statistics, there were 2,358,335 Chinese migrants in British Malaya (1940), 193,594 in Burma (1937) and 68,034 in North Borneo (1938).[80] They at the time not only had a blood and cultural link to their motherland, but many also gave their allegiance to the Guomindang regime. Since the revolutionary period of the 1890s, Chinese migrants abroad were the main financial backers and political followers of Sun Yat-sen (Sun Zhongshan 孫中山), the Guomindang's founder, and his revolutionary activities in China. The Nationalists also maintained a loose but widespread party or governmental system among Chinese expatriates, including those living in British Southeast Asia.[81] During the Sino-Japanese War, patriotism was strong among these Chinese communities. By the end of 1944, the Chinese government had received a total of CN\$ 738,341,331 in direct donations of overseas Chinese, mainly from Southeast Asia. The overseas Chinese communities also contributed heavily to the purchase of Chinese war bonds, war planes, trucks, medical supplies and other materials; and remittances to their relatives in China played an important role in China's national receipts which amounted

---

77 TNA, CO 129/588/24, telegram from Sir H. Seymour to Foreign Office, December 15, 1942.
78 TNA, CO 129/588/24, telegram from Foreign Office to Chongqing, December 18, 1942.
79 TNA, CO 129/588/24, telegram from Chongqing to Foreign Office, December 25,1942.
80 Chinese Ministry of Information, ed., *China Handbook 1937–1945: A Comprehensive Survey of Major Developments in China in Eight Years of War* (New York: The Macmillan Company, 1947), 31.
81 The party and the governmental systems were respectively under the guidance and governance of the Overseas Affairs Department of the Guomindang and the Overseas Chinese Affairs Commission (OCAC) of the Executive Yuan. Although they operated separately in principle, they usually shared overseas branches and core personals for financial saving.

to CN$ 6,153,978,000 in total from 1937 to 1944.[82] All this raised British anxiety about the allegiance of these Chinese communities, since they were not just substantial in number but also dominated the economies of Southeast Asia. London thus deemed it necessary to insert a principle of master nationality or official nationality into the treaty. Its aim was to sever the legal link between the Chinese government and these Chinese immigrants since "the Chinese have shown some disposition to ignore this rule [master or official nationality]."[83]

As far as China was concerned, however, the concept of dual nationality was introduced by the Qing government in 1909, and later inherited by the Chinese Republican and Nationalist governments, on the grounds of a traditional Chinese understanding about what constituted being Chinese, namely the Chinese citizenship was based on *Jus sanguinis*. According to Chinese culture, it was a duty for any legitimate regime to take care of its "children" living away from their homeland. Chinese migrants, no matter where they lived, would thus always be considered Chinese offspring, with their roots in China proper. Any move away from this would also be in contradiction with the Chinese nationality laws of 1929. The Chongqing government thus rejected this British proposal and London parked the issue.[84]

### 3.3.2 A National Treatment for Commerce

"Trade and commerce both with and in China", Aron Shai once remarked, "were the paramount concern at both the official and the non-official levels [for Britain]."[85] The British worried that after the abolition of extraterritoriality, their interests in China would be vulnerable to possible Chinese discriminatory laws. As a solution, London pursued a national treatment of British commercial activities in China. Under this system, both Britain and China would accord to nationals and companies of the other country treatment no less favourable than that accorded to its own nationals and companies.

To secure such a treatment for their commerce in China was desirable to both Britain and the US but during the negotiations the United States retreated from this stance since it was unable to offer the same treatment owing to its laws at state and federal levels.[86] Clarke was thus pessimistic about Sino-British negotiations over the matter:

---

82 Chinese Ministry of Information, ed., *China Handbook 1937–1945*, 33.
83 TNA, CO 129/588/24, telegram from Foreign Office to Chongqing, December 9, 1942.
84 TNA, CO 129/588/24, telegram from Sir H. Seymour to Foreign Office, December 20, 1942.
85 Shai, *Britain and China, 1941–47*, 16.
86 TNA, CO 129/588/24, telegram from Foreign Office to Chongqing, November 30, 1942.

we may not get either [a national treatment for commerce or safeguards over real property in China] since the United States government is not asking for them. If we do not get these rights in China, our only remedy will be to deny them to the Chinese.[87]

The feedback from Chongqing verified Clarke's concerns. "As the Americans have agreed to drop these words [about a national treatment of commerce] for which they had strongly pressed", the Chinese representatives remarked, "it is impossible for the Chinese government to agree to their insertion in our treaty."[88] Seymour asked to write such a treatment up as a general principle in an agreed minute, but Song Ziwen only personally and verbally assented to returning to the question in subsequent commercial treaty negotiations.[89]

### 3.3.3 Real Property

As part of the draft treaty, two International Settlements (Shanghai and Gulangyu) and two British Concessions (Tianjin and Guangzhou) were due to be abolished. This added uncertainties to the ownership of British lands and buildings and, more importantly, to whether the British could acquire new real properties in China in view of the fact that such rights were underpinned by the treaty-port system.[90] London was keen to retain these rights by reciprocating similar rights to the Chinese, even though it would be hard to swallow this in such overseas territories as Kenya. In Clarke's eyes, "the right to acquire real property in China is of great importance to British interests", but he acknowledged that "in the absence of American support it is doubtful whether we shall be successful."[91]

The Chinese Foreign Minister was at first averse to the matter. Yet the Chinese government eventually agreed to a British suggestion for compromise:

It is understood that the abolition of the system of the Treaty Ports will not affect existing property rights, and that the nationals of each High Contracting Party will enjoy the right to acquire and hold real property throughout the territories of the other High Contracting

---

**87** TNA, CO 129/588/24, letter from Ashley Clarke (FO) to W.B.L. Monson (CO), December 5, 1942.
**88** TNA, CO 129/588/24, telegram from Sir H. Seymour to Foreign Office, December 25, 1942.
**89** TNA, CO 129/588/24, telegram from Sir H. Seymour to Foreign Office, December 15, 1942 and telegram from Sir H. Seymour to Foreign Office, December 25, 1942.
**90** Bickers, *Britain in China*, 123–124.
**91** TNA, CO 129/588/24, letter from Ashley Clarke (FO) to G.F. Seel (CO), December 3, 1942. The case of Burma was similar. The Burma government had insisted upon being excluded from the formula regarding real property but at last conceded it under the Foreign Office's pressure. See TNA, CO 129/588/24, letter from Ashley Clarke (FO) to W.T. Annan (BO), December 21, 1942 and telegram from Foreign Office to Chongqing, December 27, 1942.

Party in accordance with the conditions and requirements prescribed in the laws and regulations of that High Contracting Party.[92]

In return, however, Britain dropped its claim for a national treatment for commerce.[93]

## 3.4 The Right of Coastal Trade and Inland Navigation

The question of coastal trade and inland navigation was a case outside the scope of extraterritoriality but caused intense debate between China and Britain. As discussed above, with the opening of China's inland waters, Britain accumulated considerable interests in the Chinese shipping business and did not wish to give these up. At first, London tried to turn down any Chinese demands by underlining that the rights of coastal trade and inland navigation had no connection with extraterritoriality. However, Washington disregarded this and indicated its willingness to listen.[94] Whitehall then attempted to argue that the rights concerning coastal trade and inland navigation were reciprocal and thus unreasonable to relinquish. In their eyes, Chinese vessels enjoyed the right to sail across coastal and inland waters of the United Kingdom, India and other British colonies. An illustration of this reciprocity was the fact that Chinese shipping was permitted to sail along the Irrawaddy River in Burma under Article 12 of the Anglo-Chinese Convention of 1894.[95] The Chinese, however, denied such a connection. They contended that Chinese navigation rights on the river were rationalised by the common law of the international river but not the 1894 Convention. From the Chinese viewpoint, the Irrawaddy River was an international river on the grounds that its upper stream areas, not yet demarcated between China and Britain, were claimed by the Chinese government.[96] The British attempt to demonstrate reciprocity thus proved fruitless.

---

92 The text was quoted from Clause 3 in the Exchange of Note, available from UK Treaties Online, "Treaty between His Majesty in respect of the United Kingdom and India and His Excellency the President of the National Government of the Republic of China for the Relinquishment of Extra-Territorial Rights in China and the Regulation of Related Matters (with Exchange of Notes and Agreed Minute)".

93 TNA, CO 129/588/22, "cypher telegram from Secretary of State to Government of India, External Affair Department", December 28, 1942 and telegram from Dominion Office to Canada, Australia, New Zealand and South Africa, January 7, 1943.

94 TNA, CO 129/588/24, "message from Department of State", December 1, 1942.

95 TNA, CO 129/588/24, telegram from Foreign Office to Chongqing, November 26, 1942.

96 TNA, CO 129/588/24, telegram from Sir H. Seymour to Foreign Office, December 20, 1942.

Meanwhile, without reference to Britain, the American government found it-self in a position to give up its rights relating to coastal trade and inland navigation.[97] As a safeguard, Washington merely required MFN treatment on the matter.[98] The reasons for the American compromise over this derived from domestic concerns. Foreign vessels were banned from undertaking commercial activities in American territorial waters.[99] Chiang grudgingly accepted this formula for the sake of ensuring American military assistance in the ongoing conflict against the Japanese.[100]

From the perspective of the British, the American attitude undermined their position. London now sought to give up its rights relating to inland navigation in exchange for those in regard to coastal trade.[101] But in the end there was no alternative but to fall into line with the Sino-American text. It was thus agreed that

(i) The British government relinquishes the special rights with regard to coasting trade and inland navigation in the waters of China and the Chinese government is prepared to take over any British properties which have been used for the purposes of these trades;

(ii) The Chinese government relinquishes the special rights in respect of navigation on the River Irrawaddy under Article 12 of the Convention signed at London on the 1st March 1894;

(iii) Britain and China accorded each other MFN treatment on coasting trade or inland navigation if either of these was opened to any third country;

(iv) Coasting trade and inland navigation are exempted from the requirement of national treatment.[102]

While these negotiations were proceeding Britain and China reached an impasse over the early return of the New Territories to China, a leased territory of Hong

---

**97** TNA, CO 129/588/24, telegram from Foreign Office to Chongqing, December 12, 1942.

**98** About the content of the Sino-American treaty of 1943, please see Appendix II, Fishel, *The End of Extraterritoriality in China*, 233–240.

**99** TNA, FO 371/63279, letter from G.V. Kitson (FO) to Sir Ralph Skrine Stevenson (Nanjing), February 14, 1947.

**100** December 14, 1942, see Sulan Gao, ed. *Jiang Zhongzheng zongtong dang'an: shilüe gaoben [Draft Papers of Chiang Kai-shek]*, 82 vols., vol. 52 (Taipei: Academia Historica, 2011), 62–64 (hereafter cited as *SLGB vol. 52*).

**101** TNA, CO 129/588/24, telegram from Dominion Office to Canada, Australia, New Zealand and South Africa, December 4, 1942.

**102** Quoted and slightly revised from Annex 1 (g) in Exchange of Notes, available from UK Treaties Online, "Treaty between His Majesty in respect of the United Kingdom and India and His Excellency the President of the National Government of the Republic of China for the Relinquishment of Extra-Territorial Rights in China and the Regulation of Related Matters (with Exchange of Notes and Agreed Minute)".

Kong which most Nationalist officials, including Chiang Kai-shek, regarded as being in the same category as settlements and concessions. This almost resulted in the termination of the entire Sino-British negotiation process (discussed in Chapter III). After a compromise had been accepted by the Chinese, Chongqing and London finally signed the treaty on January 11, 1943, simultaneously with its Sino-American counterpart, with an agreement to discuss the future of the New Territories after the war.

## Conclusion

The start of the Pacific War and China's alliance with the United States and Britain considerably accelerated the termination of foreign imperialism in China. Chiang Kai-shek's strategy of playing an American card in Sino-British relations in many ways worked effectively. Washington forced the British to come to the negotiating table in 1942 rather than discussing the question at the end of the war. Further, Washington gave up the original Anglo-American position of limiting the talks to extraterritorial rights in order to meet the Chinese demands. As a result, London was obliged to widen the remit of the talks to include most of its rights in China.

On the surface, it appeared that the treaty was an important step in improving Sino-British relations, but the two sides were in fact left frustrated and resentful. Chiang Kai-shek deemed the negotiation a failure due to China's inability to restore the New Territories of Hong Kong. At the same time, the British were forced to concede more than they had intended and were angered by Chiang's intransigence.

In many ways, the 1943 treaty negotiations reflected the complexity of the Sino-British-American relationship at this period of the Pacific War. The White House was ready to upset the British in order to appease Chinese demands and stimulate China's resistance against the Japanese. In this regard, the Anglo-American "special relationship" was clearly strained, at least on Chinese issues. As regards the alliance between Britain and China, their cooperation, to a large degree, was regulated by their need to placate the United States, rather than any great convergence of interests. In turn, Washington sought to maintain the Sino-British unity because it needed both Britain and China in order to win the global war.

Despite these realities, the 1943 Sino-British treaty substantively changed the landscape of the Sino-British relationship, which indicated a more equal era. After annulling most of its privileges in China, British influence over China was much reduced. This, on the one hand, enhanced the status of Hong Kong.

The city was not only an imperial territory, but a unique foothold for Britain to restore its commercial interests within China after the abolition. Therefore, the need to keep Hong Kong under the Union Jack became a key objective for Britain's China policy (discussed in Chapters III and IV). On the other hand, further concessions from the British side became extremely hard, which partly contributed to the failure of the Sino-British commercial treaty in 1943–1948. The next chapter will examine this question.

# Chapter II
# The Abortive Sino-British Commercial Treaty, 1943–1948

The negotiations over a Sino-British "comprehensive" – or commercial – treaty were a continuation of the 1943 treaty discussions. Article 8 (i) of the 1943 treaty stated that China and Britain

> will enter into negotiations for the conclusion of a comprehensive modern treaty or treaties of friendship, commerce, navigation and consular rights upon the request of either of them or in any case within six months after the cessation of the hostilities in the war against the common enemies in which they are both now engaged.[1]

The "comprehensive" treaty, despite comprising four major elements – friendship, commerce, navigation and consular rights – was conveniently termed a commercial treaty due to the overwhelming importance of commerce. Here, the two terms are used without distinction unless a special emphasis was placed on the nomenclature. The comprehensive treaty was aimed at redesigning the Sino-British relationship – particularly its commercial aspect. In this sense, the 1943 treaty was intended to be just the starting point for the building of future Sino-Britain relations; the commercial treaty would be an end point. However, while China and the US quickly came to terms in 1946, the Sino-British negotiations failed to reach an agreement and were aborted in 1948. As a result, a post-war commercial settlement was never reached. Previous scholarly works have not paid much attention to the Sino-British negotiations in the mid-1940s. The only systematic research has been undertaken by Feng Lin 馮琳, a Chinese historian, in a conference paper. He ascribes the failure to the passive attitude of the Nationalist government.[2] But, as this chapter will suggest, the ob-

---

1 Quoted from Article 8 (i) in Exchange of Notes, available from UK Treaties Online, "Treaty between His Majesty in respect of the United Kingdom and India and His Excellency the President of the National Government of the Republic of China for the Relinquishment of Extra-Territorial Rights in China and the Regulation of Related Matters (with Exchange of Notes and Agreed Minute)".

2 Lin Feng, "Zhanhou Zhong-Ying shangyue liuchan lunxi [The Study and Analaysis of the Abortion of the Sino-British Commercial Treaty after WWII]," in *The 2006 Youth Academic Symposium of the Institute of Modern History, the Chinese Academy of Social Sciences*, ed. the Institute of Modern History CASS (Beijing: Social Sciences Academic Press, 2007), 494–518. About the negotiations of the Sino-American counterpart, C.X. George Wei provided a systematic examina-

https://doi.org/10.1515/9783110706659-009

structions to achieving a Sino-British commercial treaty mainly derived from the British side.

The chapter begins with a discussion of Sino-British preparations for and preliminary discussions over a treaty. It explores the difficulties Britain faced in preparing a treaty draft, conflicting departmental interests inside Whitehall and the influence of the completion of the Sino-American commercial treaty on its Sino-British counterpart. It further investigates the divergence between the Chinese and British treaty drafts and the British decision to retreat from the treaty negotiations.

## 1 China's Eagerness for a New Commercial Treaty with Britain and the United States

Song Ziwen was eager for a quick commercial treaty with Britain and the United States. In January 1943, shortly after the formal signature of the Sino-British treaty, he proposed to the British that commercial treaty negotiations should commence in the near future.[3] The same suggestion was simultaneously delivered to the Americans.[4] Aware that the negotiations over such a treaty would require "a good deal of time", Song also advocated early preparation of the respective drafts.[5] The responses of the United States and Britain were divergent. London did not favour this proposal. Since the future world, to British eyes, was so fraught with uncertainties, any hurried and ill-considered arrangement with the Chinese might conflict with other international settlements of the same kind.[6] In contrast, the Americans felt that if such negotiations were conducted at a time when the Chinese still needed military support from the United States, they could be beneficial to US interests in China.[7] Furthermore, they saw a Sino-American treaty as analogous to treaties with other countries. In this vein, the

---

tion. See C.X. George Wei, *Sino-American Economic Relations, 1944–1949* (Westport, Connecticut and London: Greenwood Press, 1997), 85–118.

3  TNA, FO 371/35809, telegram from Chongqing to Foreign Office, January 25, 1943.

4  Memorandum of Conversation, by the Chief of the Division of Far Eastern Affairs (Hamilton), February 26, 1943, US Department of State, ed., *Foreign Relations of the United States Diplomatic Papers, 1943, China* (Washington, D.C.: Government Printing Office, 1957), 710 (hereafter *FRUS: 1943 China*).

5  Ibid.

6  TNA, FO 371/35809, letter from Ronald Fraser (Board of Trade) to Philip Broad (Foreign Office), February 25, 1943.

7  TNA, FO 371/35810, minute by Ashley Clarke, titled "Commercial Treaties with China", October 12, 1943.

treaty would be in conformity with the principles of any future global commercial framework.[8]

The attitude of the Americans placed British officials in a dilemma. The memory of British concessions forced by the Sino-American accord in 1942 was still fresh. The Foreign Office feared that the British would once again be obliged to "follow an American model" which would prejudice its important investments in China.[9] For this reason, some attempts were made to dissuade the Americans from any independent negotiations with the Guomindang, but with no success.[10] Washington merely agreed to a full information exchange with the British.[11] As a result, the United States and Britain proceeded with their preparatory work for the treaty negotiations separately.

## 2 Britain's Difficulties in Preparing a Treaty Draft, 1943–1946

The Foreign Office and the Board of Trade were the main drivers behind British commercial policies relating to China. Yet in contrast to the Foreign Office, which was more sensitive to the American angle, the Board of Trade tended to consider the treaty from an exclusively British perspective. After consulting with other governmental bodies in London and overseas, it came to the conclusion that it was unwise at this time to initiate commercial talks with China, not only due to the unclear global economic future but also because of doubts that the present Chinese government might be "unseated" at the end of the war. As one official noted, "if we are going to negotiate a treaty of such importance as this one, we would like to be sure that we are negotiating with a government whose acts would receive the approval of their successors."[12] Nonetheless, in order to dilute US influence, the Board of Trade did agree to discuss two aspects of the treaty with the Chinese: the general clauses (those for instance, relating to establish-

---

**8** Memorandum of Conversation, by Mr. Woodbury Willoughby of the Division of Commercial Policy and Agreements, May 7, 1943, US Department of State, ed., *FRUS: 1943 China*, 712–713.
**9** TNA, FO 371/35810, letter from Ashley Clarke (FO) to Ronald Fraser (BT), August 19, 1943.
**10** TNA, FO 371/35810, telegram from Foreign Office to Washington (Viscount Halifax), April 12, 1943.
**11** TNA, FO 371/35810, telegram from Washington (Viscount Halifax) to Foreign Office, May 6, 1943.
**12** TNA, FO 371/35809, letter from Mr Fraser (BT) to Philip Broad (FO), February 25, 1943.

ment, acquired rights, industrial property and unfair competition) and the navigation clauses.[13]

The British authorities and the Chinese Nationalists had previously discussed a new bilateral commercial treaty in 1929. Then, extraterritoriality had been the main obstacle to any effective commercial talks. Now, the relinquishment of extraterritorial rights had removed that barrier. That explains why Song Ziwen promptly raised the matter after the conclusion of the 1943 treaty. Yet the commercial treaty draft of 1929 was no longer suitable as a basis for Sino-British discussion.[14] The British thus had to prepare a fresh draft. Concurrently, it was decided to prepare a new consular convention which, together with a commercial treaty, would constitute a fuller readjustment of Sino-British relations for the post-war era.[15]

The removal of extraterritoriality did not make commercial treaty formulation easier. The British wished to safeguard their economic interests in China but any measures that contradicted the principle of reciprocity were unlikely to gain the Chinese government's approval. Furthermore, it became more and more difficult to form a united voice across the British Empire as a consequence of its increasingly fragmented nature. These factors contributed to the difficult birth of a British treaty blueprint.

## 2.1 A Broader or Narrower Territorial Scope?

The commercial treaty with China was one of the first such bilateral treaties that the British negotiated during the war.[16] British officials had initially to figure out to what extent any such treaty would cover the British Empire. This should not have been a difficult problem, since the 1943 treaty had already provided a way forward, namely including all British territories except the self-governing Dominions. But the Colonial Office, dissatisfied with the imperial ramifications of the 1943 treaty, took an abrasive stance. It wanted Chinese treaty rights to be merely applied to a limited number of British domains while conversely all British subjects and British-protected persons, no matter whether they were in the United Kingdom, the colonies, Dominions or British mandated or protected lands, would enjoy the benefits of the treaty.[17] Clearly, such an asymmetric arrange-

---

13 Ibid.
14 TNA, FO 371/35809, letter from Ashley Clarke (FO) to A.R. Fraser (BT), January 20, 1943.
15 TNA, FO 371/35809, letter from Ashley Clarke (FO) to A.R. Fraser (BT), February 2, 1943.
16 In 1945, the commercial treaty with China started to be jointly considered with that of Egypt.
17 TNA, FO 371/46223, letter from Board of Trade to Colonial Secretary, November 2, 1945.

ment would not be accepted by the Nationalists. From the Chinese viewpoint, the commercial treaty was an extension of the 1943 treaty, and the essence of the former had to be in conformity with the latter. Fully aware of Chinese concerns over any unequal or non-reciprocal proposals, the Foreign Office and the Board of Trade attempted to dissuade the Colonial Office.[18] However, the Colonial Office continued to press this point until 1945.[19]

Self-governing Dominions like Canada, Australia, New Zealand and South Africa were able to negotiate their own respective treaties with China.[20] When it came to Ireland, the situation was more complicated. Though London still regarded Eire as a member of the British Commonwealth of Nations, the Irish government was reluctant to admit this.[21] The cases of India and Burma were even more complex. The Indian government was averse to a commercial treaty.[22] Its main anxiety derived from the issue of Tibet, whose current commercial relations with India were regulated by bilateral treaties without Chinese involvement. Should a Sino-British commercial treaty contain the gist of the 1914 Simla convention over the status of Tibet, the Indian government would become a party to the treaty.[23] A further complication was that New Delhi did not maintain any consulates of its own in China and without a consular convention the Indians in China would have no diplomatic protection.[24] But the Foreign Office was reluctant to agree to India's partial involvement in the new framework.[25] In the end, India agreed to quit the Sino-British negotiations.[26] The Sino-British commercial settlement did not captivate the Burma Office either.[27] The Burma Office

---

**18** TNA, FO 371/41601, memorandum by Board of Trade, titled "Territorial Scope of the Proposed United Kingdom-China Commercial Treaty", January 1944 (no specific date).
**19** TNA, FO 371/46223, letter from G.H. Hall (CO) to Sir Stafford Cripps (BT), November 13, 1945.
**20** TNA, FO 371/46220, telegram from New Zealand to Dominions Office, May 30, 1945 and telegram from Canada to Dominions Office, June 7, 1945; TNA, FO 371/46221, telegram from Australia to Dominions Office, July 2, 1945; TNA, FO 371/46222, telegram from Union of South Africa to Dominions Office, August 14, 1945.
**21** According to a public statement of December 30, 1937, the British government still regarded Ireland as a member of the British Commonwealth of Nations. But there was some doubt as to whether the Eire government regarded itself as constituting part of the British Commonwealth. See TNA, FO 371/46223, letter from Mr Kimber (DO) to W.E. Beckett (FO), October 29, 1945.
**22** TNA, FO 371/35809, telegram from the Indian government to India Office, March 26, 1943.
**23** TNA, FO 371/46221, letter from A. Dibdin (IO) to J.C. Sterndale Bennet (FO), July 14, 1945. Also see Chapter V.
**24** TNA, FO 371/46220, letter from R.T. Peel (IO) to J. Sterndale Bennett (FO), March 9, 1945.
**25** TNA, FO 371/46220, letter from A.L. Scott (FO) to D.M. Cleary (IO), May 11, 1945.
**26** TNA, FO 371/46222, telegram from External Affairs Department (GOI) to India Office, August 11, 1945.
**27** TNA, FO 371/41601, letter from W.T. Annan (BO) to Ashley Clarke (FO), March 28, 1944.

was apprehensive about the likely influx of Chinese immigrants, which had been on a massive scale during the war. Its desire was to review all existing bilateral treaties between China and Britain, and in particular to abrogate Article 14 of the 1894 Convention, which put no restriction on Chinese migration.[28] Yet, as far as the Foreign Office was concerned, this carried a serious risk to British treaty rights regarding Hong Kong. If, for example, Article 2 of the Nanjing Treaty (in reference to the opening of five treaty ports) was deemed invalid, why was Article 3 in the same treaty, which ceded Hong Kong Island to Britain, still in force?[29] Besides the concern of immigration, the Burma Office was also worried about conceding mining and land rights to Chinese merchants, which were part of the British strategy to please the Chinese so that the latter could reciprocate by offering the same rights to the British.[30] Burma's opposition became stronger when a more independent Burmese government emerged after the war. Concessions over immigration and land ownership to China were deemed detrimental to Burma's interests.[31] As a result, the Burma regime refused to be entangled by any Sino-British stipulations.[32]

Further opposition came from the British colonies of Malaya and Hong Kong. Malaya was concerned about Chinese immigration and the Colonial Office was disposed to let Malaya decide the matter itself given the issue was an "extremely delicate one".[33] Yet the Foreign Office declined to give British Malaya control over immigration because it would be interpreted as London's view and difficult to separate from discrimination against the Chinese.[34] However, the newly established Malayan Union would not accept this.[35] For his part, the governor of Hong Kong was also reluctant to sacrifice the colony's interests in a Sino-British

---

**28** TNA, FO 371/46220, letter from W.T. Annan (BO) to J.C. Sterndale Bennett (FO), January 17, 1945.
**29** TNA, FO 371/46221, letter from Mr. Willis (BT) to J.C. Sterndale Bennett (FO), July 28, 1945.
**30** TNA, FO 371/46222, letter from Mr. Willis (BT) to L.B. Walch Atkins (BO), September 28, 1945.
**31** TNA, FO 371/53659, telegram from W.R. Bickford (Secretary of Defence and External Affairs Department in the Burma government) to the Under-Secretary of Burma Office, May 13, 1946.
**32** TNA, FO 371/63280, letter from Mr. More (BO) to A.L. Scott (FO), June 25, 1947.
**33** TNA, FO 371/46222, letter from Mr. Davies (CO) to J.R. Willis (BT), August 28, 1945.
**34** TNA, FO 371/46222, letter from G.V. Kitson (FO) to J.R. Willis (BT), September 5, 1945.
**35** TNA, FO 371/53659, letter from Sir E. Gent (Malayan Union) to Secretary of State for Colonial Office, October 18, 1946; TNA, FO 371/69625, letter from G. Darby (CO) to H.O. Hooper (BT), November 20, 1948. The Malayan Union, consisting of 11 states, was a single British Crown colony imposed by the British during the period between 1946 and 1948. Due to the opposition of the Malayan nationalists, it disbanded in 1948 and was replaced with the Federation of Malaya, which restored the symbolic authority of the rulers of the Malayan states. In 1957 the Federation became independent and was reconstituted as Malaysia in 1963 after incorporating Sarawak, North Borneo and Singapore. Singapore left it and became an independent republic in 1965.

treaty. Such a treaty was likely to involve migration control in the colony, in conflict with freedom of entry that was traditionally enjoyed by people of the Chinese race; reciprocal provisions for insurance companies which might have a detrimental effect on local businesses; the curtailment of the existing powers of deportation which could have an impact on law and order; and the omission of regulations about the safety of life at sea.[36] But Hong Kong, as a tiny British Crown colony, had no authority to override London.[37] The colony was thus dragged into the treaty involuntarily.

Meanwhile, in contrast to the British, the Americans initiated exploratory work as early as possible.[38] This progressed smoothly. By May 1944, a tentative US draft had already been completed.[39] The Chinese draft was also nearly ready. The two countries agreed to start negotiations within six months after the end of hostilities.[40] In practice, the timing was even quicker. Treaty drafts had been swapped by April 1945.[41] However, despite promising to do so, Washington refused to share details of the draft with London. The US State Department wished to avoid being accused of any Anglo-American collusions against the Chinese.[42] This made London nervous.

## 2.2 No Clauses about the Treatment of Goods

The speed with which the US acted created tensions in Whitehall. The Foreign Office longed to catch up with the Americans on this matter. But the Board of Trade advocated dealing with the matter at London's own pace.[43] Furthermore,

---

36 TNA, FO 371/53659, telegram from G. Darby (CO) to J.R. Willis (BT), September 19, 1946 and telegram from Mark Young, Governor of Hong Kong, to G. Darby (CO), August 22, 1946. The British draft had no stipulations about the safety of life at sea. But the matter had been covered by the Draft Treaty of the International Convention for the Safety of Life at Sea, 1929, and the International Load Line Convention, 1930, both of which China was a signatory.
37 TNA, FO 371/53660, telegram from the Secretary of State for the Colonies to the Governor of Hong Kong, December 19, 1946.
38 TNA, FO 371/35810, telegram from Washington (Viscount Halifax) to Foreign Office, May 6, 1943.
39 TNA, FO 371/41601, letter from Ashley Clarke (FO) to A.R. Fraser (BT), May 17, 1944.
40 TNA, FO 371/41601, telegram from Chancery (Chongqing) to Far Easter Department (FO), June 28, 1944.
41 TNA, FO 371/46220, telegram from Sir H. Seymour (Chongqing) to Foreign Office, April 17, 1945.
42 TNA, FO 371/46220, telegram from Earl of Halifax (Washington) to Foreign Office, May 29, 1945.
43 TNA, FO 371/41601, letter from Mr. Welch (BT) to Ashley Clarke (FO), June 28, 1944.

since the British had already agreed a tariff treaty with China in 1928, which covered bilateral tariff terms, and there was anxiety about possible clashes with future international economic policies, the Board of Trade's preference was in favour of shelving any clauses regarding to the treatment of goods.[44] The Foreign Office doubted whether the Chinese would accept this, considering that they generally regarded the 1928 treaty as "a relic of the unequal treaties". It was also concerned that British goods might be left without any safeguards in China – with potentially serious damage to the British post-war economic recovery, and that the Chinese might frown upon a treaty without goods clauses as not comprehensive.[45]

The arguments of the Foreign Office, however, swayed neither the Board of Trade nor the Treasury. Although it drafted a set of clauses dealing with all aspects of the subject (e. g. duties, import restrictions and state trading), the Board had little interest in encouraging British exports to China since China was not a hard currency market. Difficulties in the balance of payments further hindered Britain from reciprocating any MFN status to China relating to import restrictions.[46] As far as the Treasury was concerned, some safeguards contemplated by the Board of Trade for the treaty were not only "quite useless" but also created new trouble. The introduction of import restrictions ("Allocation of Import Quotas") would do little to relax Chinese control over imports since "the Chinese will always say that they have an inadequate supply of exchange." In contrast, implementing the same policy in Britain would tie British hands unless Britain had the same excuse of an inadequate supply of Chinese dollars.[47] Similarly, any restrictions on a trade being monopolised by the state, the Treasury believed, would be futile. It "should not succeed in getting the Chinese to observe them in a way which would preserve our own legitimate interests."[48] Eventually, a compromise emerged between the Foreign Office on the one hand and the Board of Trade and the Treasury on the other. Articles about the treatment of goods were to be absent from any treaty discussed with the Chinese authorities

---

**44** TNA, FO 371/41601, letter from Mr. Stirling (BT) to W.G. Weston (Ministry of War Transport), August 12, 1944.
**45** The opinion of the Foreign Office was quoted by the Board of Trade in its document. See TNA, FO 371/46220, letter from Sir Percival Liesching (BT) to Sir Wilfrid Eady (the Treasury), March 16, 1945.
**46** TNA, FO 371/46222, letter from Mr. Welch (BT) to Under-Secretary of State (FO), October 5, 1945.
**47** TNA, FO 371/46220, the note of March 21, 1945 annexed to the letter from Sigi Waley (the Treasury) to Sir Percival Liesching (BT), March 28, 1945.
**48** Ibid.

and, as a result, only two items remained on the British agenda: the clauses regarding establishment and navigation.[49] In these circumstances, the Board of Trade emphasised the extension of the treaty to British colonies so as to increase the attractiveness of the British treaty to the Chinese.[50]

## 2.3 Relaxing British Border Controls

Two issues, in the opinion of the Board of Trade, were equally important for British commerce with China to flourish: the freedom of British nationals to reside abroad for the purpose of carrying out business; and the right of British firms abroad to employ key personnel of their choice. However, these traditional rights had been on the verge of abolition due to the removal of extraterritoriality and the rise of Chinese nationalism. As a result, the Board suggested that the government secure these rights by opening up the borders of British territories to the Chinese.[51]

The Home Office and the Ministry of Labour were averse to this. The administration of the Aliens Order in Britain was based on the principle that a foreigner had to satisfy an immigration officer that there was a good reason for their seeking to come to Britain.[52] Yet the Board of Trade's formula was to allow the Chinese to freely enter the United Kingdom for a short visit or for a longer stay if their reasons were among the "approved purposes" listed in the proposed treaty (e.g. international trade, the supervision of business interest or systematic educational study) and there was no specific objection to them. Moreover, there was the likelihood that the nationals of European countries might request similar terms to come to the UK to seek employment.[53] Although reassurances were given to the Home Office about its right to control immigration on the grounds of national security, public order, indigence, or breach of any conditions im-

---

49 TNA, FO 371/46223, telegram from Foreign Office to Sir Horace Seymour (Chongqing), November 20, 1945.
50 TNA, FO 371/46221, letter from Mr. Shackle (BT) to Sir G. Clauson (CO), June 25, 1945.
51 TNA, FO 371/46220, letter from R.J. Shackle (BT) to E.D. Carew Robinson (HO), May 3, 1945.
52 TNA, FO 371/46221, letter from O.D.C. Robinson (HO) to R.J. Shackle (BT), June 28, 1945.
53 TNA, FO 371/46223, letter from J. Chuter Ede (HO) to Sir Stafford Cripps (BT), November 23, 1945 and memorandum by the Foreign Office, the Home Office, the Board of Trade and the Ministry of Labour, titled as "Establishment Articles in Commercial Treaty", prepared for the Cabinet conference, December 6, 1945.

posed, it insisted that the matter be decided by the Cabinet.[54] At a Cabinet meeting in December 1945, the Home Office eventually fell into line with the proposal of the Board of Trade.[55]

## 2.4 Coastal Trade and Inland Navigation

London, as we saw in the previous chapter, grudgingly relinquished its special rights regarding coastal trade and inland navigation under Sino-American pressure. According to the 1943 treaty, any further opening up of Chinese markets to foreign shipping entirely depended on the Nationalist government's willingness to do so.[56] The British, however, wanted to reverse this in the commercial treaty. Before the war, Britain had been the dominant maritime power in China, owning not merely huge shipping fleets but also facilities in most ocean or river ports, such as quays, docks, warehouses and fuel stores. These interests had previously been protected by extraterritoriality and the treaty-port system. The value of such property owned by British shipping companies was somewhere between £15 and £20 million, according to the statistics of the British Ministry of Transport, and the annual profit was about ten per cent of the total value of their capital.[57] After the 1943 treaty, the Foreign Office thus found itself in a whirlpool of public criticism. In Parliament, Richard Law, the Parliamentary Under-Secretary of State for Foreign Affairs, was questioned as to "why [we should] abrogate a privilege that has been in operation for many years and thus strike a very deadly blow at our Mercantile Marine?" As an expedient, Law committed himself to recheck the matter "when the time comes for a comprehensive arrangement."[58] In the editorial of *The Syren and Shipping Illustrated*, the censure was even stronger:

---

**54** TNA, FO 371/46223, "note on the establishment articles of the draft Establishment and Navigation Treaties with China and Egypt", by Ian Wilson Young (FO), December 11, 1945, and letter from J. Chuter Ede (HO) to Sir Stafford Cripps (BT), November 23, 1945.

**55** TNA, FO 371/46223, minute by G.V. Kitson (FO), December 14, 1945.

**56** UK Treaties Online, "Treaty between His Majesty in respect of the United Kingdom and India and His Excellency the President of the National Government of the Republic of China for the Relinquishment of Extra-Territorial Rights in China and the Regulation of Related Matters (with Exchange of Notes and Agreed Minute)".

**57** TNA, FO 371/63281, note by Ministry of Transport on shipping points made by Mr. Roland Lion in his letter of October 14, 1947 to Mr Bottomley, October 29, 1947.

**58** House of Commons Debates, vol. 393, November 10, 1943, col. 1121, https://api.parliament.uk/historic-hansard/commons/1943/nov/10/coasting-trade (accessed April 2, 2019).

the Government have actually handed to the Chinese the whole of the China Coast trade which has been largely created and carried on for the past century by British shipping enterprise – and they have done so without consulting the British interests involved. ... We state frankly that we are amazed by this action on the part of the Government. It argues ill for the future of this country and is in our view a sign that this pernicious policy of appeasement has so permeated the purlieus of Westminster that all thought of upholding the position and prestige of Britain has disappeared from the Government mind. A trade that existed to a large extent because of the courage and enterprise of four British shipping companies – Alfred Holt and Co., Jardine, Matheson and Co., John Swire and Sons and the Moller interests – has been gratuitously presented to the Chinese by our Government without one word of warning to the interested parties.[59]

A strong body of opinion thus gradually developed in London to override the shipping provisions in the 1943 treaty. The Foreign Office advised that the Chinese government would not compromise on the question. In its opinion, the best path was gracefully to accept the principle that Britain had no treaty rights of navigation in coastal and inland waters, and leave the shipping companies, with normal diplomatic assistance, to obtain concessionary rights privately.[60] However, the fact that the Chinese shipping industry had recovered after the war exacerbated British anxiety that all the trade in the region was bound to be set aside exclusively for Chinese companies, private or state-owned.[61] As a result, the British government began to think about how it could regain these rights through a commercial treaty. Therefore, stipulations about the new treaty being in conformity with the principles of the 1943 treaty were removed from the British draft and at the behest of the Ministry of War Transport (which became the Ministry of Transport after the war) no differentiation was to be made between coastal and inland water trade, and ocean-going trade, and there was to be national treatment in both spheres. In other words, British vessels would enjoy the same rights as those accorded to Chinese vessels.

The Ministry of Transport also advocated a "fair play" code in the treaty. This was aimed at outlawing subsidies (except in defined circumstances) and other discriminatory practices. The code included sanctions, such as countervailing or fighting subsidies, and denying particular ships (or ships of particular

---

**59** No Author, "On Watch," *The Syren & Shipping Illustrated* 189, no. 2463 (1943): 153–156.

**60** TNA, FO 371/41601, letter from A.D. Blackburn (FO) to A.R. Fraser (BT), February 21, 1944.

**61** China's merchant shipping suffered a loss of 300,000 tons during the war and no indemnification had been effected by 1947. But the state-operated China Merchants after the war took over a large number of enemy vessels and acquired new ships. See TNA, FO 371/63280, "Shipping Industry Acquaints Government with Difficulties," *Financial Daily News* (March 14, 1947), translated and annexed to the telegram from Sir R.S. Stevenson (Nanjing) to Foreign Secretary (London), April 14, 1947.

flags) the use of shipping facilities or excluding them from trade altogether. In the light of the fact that a large number of British ships formerly trading with China were registered in Hong Kong, the Ministry was delighted to see that Hong Kong, and other such colonies, were covered by the treaty. Moreover, the treaty, it was emphasised, should not conflict with any British future effort to "[develop British] shipping interests in China indirectly through share holding and financial control of ships flying the Chinese flag."[62]

In the end, after nearly three and a half years of agonising over the details, the British treaty blueprint was formally presented to the Chinese government on June 19, 1946. In the interim, the British draft of the Consular Convention had been delivered to the Chinese in February of the same year.[63]

## 3 The Signing of the Sino-American Treaty of Friendship, Commerce and Navigation

While British diplomats aspired to win back British privileges through their own commercial treaty, its Sino-American counterpart was signed on November 4, 1946. The Sino-American treaty, directly growing out of the 1943 agreement, provided freedom for citizens, businesses and associations of one country to reside, travel and trade in all parts of the other country. Except for an American amendment reserving the right to grant special treatment in trade matters to Cuba and the Philippines, which China could not claim, the treaty was fully established on the basis of equal and reciprocal principles.[64] Compared to previous Sino-British treaties, or the British proposed treaty draft, the 1946 Sino-American settlement was on a much broader scale, including articles of establishment, landholding, industrial and literary property, commerce and international trade expansion, international payment and exchange controls, customs regulations and formalities, import and export restrictions, government monopolies and public agencies, as well as other matters.[65] Also, in contrast to the British inclination to

---

62 TNA, FO 371/41601, letter from W.G. Weston (MWT) to J.R. Willis (BT), December 20, 1944.
63 IMH Archives, 11–10–06–01–022, *Zhong-Ying shangyue* [The Sino-British Commercial Treaty], letter from British Embassy in Nanjing to the Chinese Foreign Ministry, June 19, 1946; TNA, FO 371/53659, telegram from British Embassy (Nanjing) to Foreign Office, June 29, 1946.
64 Times Correspondent, "American Treaty with China: Commerce and Navigation," *The Times*, November 5, 1946.
65 TNA, FO 371/53660, letter from Sir R. Stevenson to Mr. Attlee, November 25, 1946 and American document for the Press about the news of Sino-American treaty (November 2, 1946), transited by the British Embassy at Washington to Foreign Office, November 2, 1946.

retain old treaties, the American treaty was aimed at superseding all existing treaties relating to commerce, navigation and the entry of nationals to each other's territories, but also not limiting or restricting the rights, privileges and advantages hitherto accorded by the 1943 treaty.[66] Furthermore, the treaty embodied a breakthrough in international legal terms. For example, Article 28 stated that any dispute between the two governments as to the interpretation or application of the treaty which the governments could not adjust satisfactorily by diplomacy should be submitted to the International Court of Justice, the principal judicial organ of the United Nations established in 1945, unless the two governments could agree to settlement by some other specific means.[67]

The Sino-American treaty was interpreted by some commentators in Britain as a serious American economic threat to British interests in China. *The New Statesman and Nation* denounced the treaty as a good example of American imperialist expansion which "gives complete rights to Americans to exploit China without let or hindrance."[68] Meanwhile, MPs from both major parties urged the British government to expedite negotiations with China and develop "friendly relations between the two peoples through the extension of mutual cultural and trade interests."[69] In line with such thinking, a parliamentary goodwill mission to China was planned for the summer recess of 1947.[70] But when it came to the question of Sino-British treaty negotiations, the Sino-American agreement created several hindrances to British freedom of action.[71]

These obstacles were primarily in three fields. First, the American treaty precluded alien land ownership and merely allowed foreigners to lease land. To acquire, hold and dispose of real and other immovable property was permitted on reciprocal grounds, but, in the American legal context, various forms of laws existed in a number of US states to block foreigners or the Chinese from possessing buildings or land.[72] This gave the Chinese government a warrant to debar its real estate market from the Americans in the same manner. Second, the treaty was more concerned with overseas trade, rather than coastal trade or inland navigation, because the US banned foreign vessels from undertaking commercial activ-

---

**66** TNA, FO 371/53659, report from U.S. Information Service, November 4, 1946.
**67** Ibid.
**68** No Author, "American's Empire," *The New Statesman and Nation* 32, no. 822 (1946): 371–372.
**69** House of Commons Debates, vol. 431, December 4, 1946, cc. 339–340, https://api.parliament. uk/historic-hansard/commons/1946/dec/04/anglo-chinese-relations (accessed April 2, 2019).
**70** Ibid.; TNA, FO 371/53660, telegram from Foreign Office to Nanjing, December 14, 1946.
**71** TNA, FO 371/53659, letter from A.L. Scott (FO) to H.O. Hooper (BT), September 16, 1946.
**72** Charles J. Fox, "The Sino-American Treaty – I," *Far Eastern Survey* 16, no. 12 (1947): 139–142.

ities in its own territorial waters.[73] Thus, the United States merely agreed to enjoy the benefits of any relaxation of Chinese restrictions on cabotage and inland navigation which might be made in favour of third parties.[74] Third, the treaty defined "corporations and associations" as corporations, companies, partnerships and other associations, whether or not with limited liability and whether or not for pecuniary profit, but omitted some non-political associations: clubs, chambers of commerce, guilds, schools and hospitals, for instance. These were also omitted in Chinese Company Law, in force since April 1946, under which a body eligible to be registered as a company was only "an association for profit".[75] These definitions were a particular source of concern to British communities in China as they cast doubt over whether they could continue to establish and run such institutions in the country.[76] Furthermore, chambers of commerce, which had been the hubs of British commercial activities in China, were about to be placed under Chinese strict jurisdiction.[77]

## 3.1 Chinese Reactions to the Sino-American Treaty

The official Chinese response to the Sino-American commercial treaty was one of triumph. The Nationalist government not only regarded the signing of the treaty as a diplomatic victory after the termination of the "unequal treaties", but also as an indication of the United States' endorsement of the Nationalist regime in the midst of the civil war. The signing ceremony was accordingly upgraded to

---

**73** TNA, FO 371/63279, letter from G.V. Kitson (FO) to Sir Ralph Skrine Stevenson (Nanjing), February 14, 1947.
**74** TNA, FO 371/53660, letter from Sir R. Stevenson to Mr. Attlee, November 25, 1946.
**75** TNA, FO 371/63280, letter from Yee Tsoong Tobacco Distributors Limited (Shanghai) to J.C. Hutchison (Commercial Counsellor, British Embassy), January 16, 1947.
**76** TNA, FO 371/63280, letter from Imperial Chemical Industries (China) LTD to J.C. Hutchison (Commercial Counsellor, British Embassy), January 16, 1947.
**77** A set of Chinese "Measures Governing the Organization of Chambers of Commerce by Foreign Merchants and their participation in (Chinese) Trade Organisations" permitted the organisation of a foreign chamber of commerce if five or more foreign merchants of the same nationality were in the city. All foreign chambers of commerce had to register at the local Social Affairs Bureau. It was also hoped that foreign merchants might join trade guilds, so that they would be bound by legal agreements among members. These local guilds, from the British perspective, were masquerades for Chinese supervisory bodies. See TNA, FO 371/53660, translation of extract from *Commercial Journal* of September 30, 1946, annexed to telegram from Commercial Secretariat (Shanghai) to British Embassy (Nanjing), titled "Participation of Foreign Merchants in Chinese Trade Organisations", September 30, 1946; TNA, FO 371/53659, telegram from J.C. Hutchison (Shanghai) to Leo H. Lamb (Nanjing), August 7, 1946.

take place at the Executive Yuan rather than the Ministry of Foreign Affairs as was usual.[78] Ralph Stevenson, the British ambassador to China since May 1946, observed that "the State Department had not felt signatures advisable as it might be interpreted as encouraging the Central Government to think that American support would be forthcoming in other directions."[79] Indeed, the Truman government had experienced some difficult relations with the Nationalists in 1945–1946 over differing views as how to handle the Chinese Communist Party (CCP). After the Japanese defeat, the CCP had grown up as an impressive political entity with powerful military forces. Washington wished that the two parties should settle their differences by peaceful negotiation and form a coalition government. In this manner, they could compete with each other for the opportunity to govern China through parliament, rather than on the battlefield. With such a purpose in mind, an American diplomatic mission, headed by George Marshall, the recently retired Chief of Staff of the US Army and later the US Secretary of State (1947–1949), was present in China between December 1945 and January 1947. Yet, neither the Nationalists nor the CCP deemed a democratic government to be a practical solution to their conflict. The Marshall mission thus proved fruitless at delivering China from its dreadful domestic hostilities. To exert more pressure on the Nationalist government, Marshall, with the approval of President Harry S. Truman, imposed a weapons embargo from July 1946 to May 1947.

The response of the Chinese press to the treaty was different to that the Nationalist government. With the exception of the official *Zhongyang Ribao* 中央日報 (*Central Daily News*), it was generally opposed to the treaty. *Xinwenbao* 新聞報 pointed out that the treaty was more favourable to the United States than to China since nationals of the former country, being richer, could exercise a wider range of privileges, including the right to reside, travel, carry on trade, engage in commercial transactions, educational activities, industry, navigation and the exploitation of mineral resources.[80] Likewise, *Qianxian Ribao* 前線日報 remarked that the principles of friendship, equality and reciprocity had been

---

**78** TNA, FO 371/53695, "Synopsis of the Sino-American Commercial Treaty" issued by the London Office of the Chinese Ministry of Information, November 7, 1946.
**79** TNA, FO 371/53660, letter from Sir R. Stevenson to Mr. Attlee, November 25, 1946.
**80** These Chinese newspaper comments are drawn from a British telegram. See TNA, FO 371/53660, telegram from Sir R. Stevenson to Mr Attlee, December 12, 1946.

fully demonstrated through the treaty, but in view of the relative material strength of the two countries, there could hardly be perfect equality.[81]

*Dagongbao* 大公報 and *Wenhuibao* 文匯報 were even harsher, condemning the treaty as the first of a new set of "unequal treaties". *Dagongbao* took particular exception to the fact that most of the clauses contained a most-favoured-nation treatment stipulation and argued that this, combined with a quotation from a Chinese government source that the treaty was to serve as a blueprint for other such treaties, virtually served the same function of the Treaty of Nanjing in 1842 in opening up the Chinese market and the country's resources for foreigners to exploit. For its part, China was in practice prevented from enjoying any rights reciprocated by the treaty in American territories due to its own economic weakness.[82] *Wenhuibao* was equally aggrieved:

a. The treaty was secretly negotiated by the Chinese Government without reference to public opinion;
b. It was doubtful that the intention of the trade pact had no political overtones against the Chinese Communists, as emphasised by both the Chinese and American spokesmen;
c. The so-called 'reciprocity and equality' embodied in the Treaty could hardly be realised in the present situation. For instances, how could Chinese citizens enjoy the right to reside and travel in the United States when they would be barred by the American Immigration Law? Would the United States Government allow Chinese companies to open mines in its territory, and would there be any Chinese vessels to navigate along the American coasts?
d. The most-favoured-nation clause stipulated in the treaty was nothing but the so-called 'Open Door' and 'Equal Opportunities' policy.[83]

In addition, several public meetings, such as that organised by the "Chongqing Bureau of Industrial & Commercial Associations" on November 29, 1946, were held in Chongqing, Hankou and Shanghai to protest against the terms of the Sino-American treaty.[84] Meanwhile, there was a growing public clamour against

---

**81** TNA, FO 371/53660, "Chinese Press Comments" extracted from the China Daily Tribune (November 6, 1946), annexed to the telegram from J.C. Hutchison (Shanghai) to British Embassy (Nanjing), November 13, 1946.

**82** The Chinese newspaper comments are quoted from a British telegram. See TNA, FO 371/53660, telegram from Sir R. Stevenson to Mr Attlee, December 12, 1946.

**83** TNA, FO 371/53660, "Chinese Press Comments" extracted from the China Daily Tribune (November 6, 1946), annexed to the telegram from J.C. Hutchison (Shanghai) to British Embassy (Nanjing), November 13, 1946.

**84** TNA, FO 371/53660, telegram from A.T. Cox (Chongqing) to J.C. Hutchison (Commercial Counsellor at Shanghai), November 29, 1946.

foreign goods being "dumped" on the Chinese market. In Chongqing, there were frequent grumbles from the local cigarette manufacturers who complained about the large quantities of Philip Morris cigarettes being sold locally. The Vice-Chairman of the Chongqing Chinese Chamber of Commerce, in his speech on November 12 at a dinner given to a United Kingdom Trade Mission, stressed the damaging impact of such "dumping" of foreign goods in China:

  a. Various products of and factories in Sichuan were stated to be suffering from unfair foreign competition (the unfairness being ascribed to 'cheapness');
  b. The latest sufferer was 'native medicines' of Sichuan;
  c. Local match manufacturers were not affected by foreign competition since their factories were working full time.[85]

Alongside this, the Chongqing insurance companies held a press conference on November 29 to criticise the unfair competition of foreign insurance companies and to inaugurate a move to combat this.[86]

In the view of A.T. Cox, the Acting British Consul-General in Chongqing, "it is rather like the old game of 'blaming the foreigner' when things go wrong ... but it is only since the publication of the Sino-American Treaty that these 'complaints' have been prominent and frequent."[87] In fact, the logic underlying these criticisms and protests was similar to that behind the trade protectionism that had influenced the thinking of Chinese Nationalist officials and economists since the late 1920s.[88] From 1927 to 1937 it was Nationalist tariff policy to build a trade wall against the import of foreign goods so as to nurture Chinese industries, although in reality this always gave way to tariff revenue growth.[89]

These Chinese popular sentiments upset the United States. John Leighton Stuart, the American Ambassador, was obliged to explain the misconceptions in the Chinese press:

A misunderstanding of the terms of the treaty, especially in regard to the most-favoured-nation clause, has erroneously led a section of the Chinese press into a false prophecy that the Treaty is the first of a series of new unequal treaties. That is not true. The fact is that the most-favoured-nation clause is included in modern commercial treaties between Powers of equal importance. ...

---

**85** TNA, FO 371/53660, telegram from A.T. Cox (Chongqing) to J.C. Hutchison (Shanghai), December 3, 1946.
**86** Ibid.
**87** Ibid.
**88** Boecking, *No Great Wall*, 38–45.
**89** Ibid., 156–157.

The United States does not expect China to sacrifice or compromise its sovereignty and the treaty in no way provides for this. The treaty does provide, however, that Americans in China and Chinese in the United States shall [in no way] be subjected to discriminatory treatment and that each shall enjoy in the other country reasonable and necessary rights and facilities.

The United States has not sought for nor has it been accorded any special rights or privileges in China.[90]

Ironically, there was also criticism in the United States. Some aspects of the treaty were thought disadvantageous to American interests. For example, the MFN treatment was not unconditional, and the International Court of Justice was given jurisdiction over possible Sino-American disputes, which conflicted with legislation passed by US Congress. Thus, there was some doubt as to whether the United States Senate ought to ratify it without reservation.[91] It was not until November 1948 that the Senate finally passed the accord.

According to Ralph Stevenson, the hostility in the Chinese press emanated from the Left and the line taken by the opponents of the Nationalist government maintained that the "government has tied itself to the United States chariot as the price of procuring further aid against the Communist."[92] Chiang Kai-shek did not believe this to be the case. He felt that the Sino-American treaty was "a perfectly normal treaty on a fully reciprocal basis and yet the Communist propaganda on the subject had met with enough success to make things very awkward for the Government."[93] Assessing whether the CCP was indeed behind the press coverage and protest marches requires further research. Nonetheless, the negative Chinese public attitude towards the Sino-American treaty made the Nationalist regime even more determined to resist certain British claims. As one British source confirmed, the Nationalist government would be more cautious and stricter during the Sino-British discussions.[94]

---

90 TNA, FO 371/53660, document of "Dr Stuart Explains new Sino-American Treaty", November 25, 1946.

91 No Author, "Commercial Treaty Delay," *Shanghai Evening Post*, vol. 66, no. 155 (July 7, 1947). Quoted from TNA, FO 371/63281, attachment to telegram from L.H. Lamb to China Department (FO), September 15, 1947.

92 TNA, FO 371/53660, telegram from Sir R. Stevenson to Mr. Attlee, December 12, 1946.

93 Chiang's words were part of his conversation with Sir R. Stevenson, which were relayed by Stevenson and recorded in one of his telegrams to London. See TNA, MT 59/2646, telegram from Sir R. Stevenson to M.E. Dening (FO), January 29, 1947.

94 TNA, FO 371/53660, telegram from British Embassy (Nanjing) to China Department (FO), December 4, 1946.

## 4 The Gap between China's and Britain's Treaty Proposals

Chinese counter-proposals to those of the British were drawn up quickly and presented to London in January 1947.[95] These were based on the following principles: a Sino-British treaty should be a full-scale scheme, comprised of three sections: friendship, commerce and navigation, and superseding all former commercial agreements between China and Britain; a Sino-British treaty should not conflict with any Chinese domestic legislation; and Chinese concessions should not go beyond the limits defined by the Sino-American treaty of 1946 and the Sino-British treaty of 1943, unless those provisions were more beneficial to Chinese interests.[96] In addition, the Chinese wanted to inventory the previous treaties superseded or nullified by the new commercial treaty.[97] However, not only were the Chinese principles a long way from the British objective of giving adequate protection to British business interests inside China, but also Whitehall believed that any attempt to agree a list of superseded treaties with the Chinese would almost inevitably lead to embarrassing political controversy, for instance, in relation to Hong Kong.[98] Unsurprisingly, the Chinese counter-draft was deemed "unsatisfactory" from the British perspective.[99]

### 4.1 The Rights of Nationals and Companies

The provisions concerning national citizen rights caused comparatively little friction between China and Britain. The British were eager to maximise the freedom of British nationals to enter and leave China for the purpose of trade and for certain professional and business activities. The Chinese motive for going along

---

**95** TNA, FO371/63279, telegram from Sir R. Stevenson to Mr. Bevin, January 13, 1947. An English version of the Chinese draft was also annexed to the telegram.

**96** IMH Archives, 11–10–06–01–022, *Zhong-Ying shangyue* [The Sino-British Commercial Treaty], minutes of the first conference examining the British treaty draft, November 11, 1946.

**97** IMH Archives, 11–10–06–01–024, *Zhong-Ying shangwuxieyue juan* [The Sino-British Commercial Agreements], minute by the Foreign Ministry about the substitution of the commercial treaty for previous treaties with Britain, January 16, 1947.

**98** TNA, FO 371/63279, Summary of main points of difference between the United Kingdom Draft Treaty of Establishment and Navigation as presented to the Chinese government in June 1946 and the Chinese Counter Draft Treaty of Friendship, Commerce and Navigation, as presented to His Majesty's Representative in Nanjing in December 1946, note by the Commercial Relations and Treaties Department of the Board of Trade, February 20, 1947 (hereafter "Summary of main points" by the Board of Trade, February 20, 1947).

**99** Ibid.

with this was not an interest in promoting Chinese commerce in the UK, or welcoming British competition in China. Their incentive was to safeguard the right of Chinese immigrants to freely enter or leave the British-dominated territories in South and Southeast Asia: Hong Kong, Singapore, Malaya, North Borneo, Burma and India. This would partially contribute to the return of Chinese immigrants to these areas who had fled to China during the war, and help them to restore properties in their name.[100]

For similar reasons, the Guomindang authorities, especially the Overseas Chinese Affairs Commission (OCAC) , were disappointed about the narrow territorial scope *ab initio* referred to in the British draft. It was one of the Nationalist core concerns to ensure that the treaty covered the same territories as the 1943 treaty, including all British domains except self-governing Dominions.[101] In particular, the new treaty should not impair the existing treaty rights held by Chinese nationals to enter and/or reside in Hong Kong and Burma.[102] However, due to the independence movements in India, Burma and Malaya, it was no longer straightforward for London to comply with this.

The matter of dual nationality was also relatively uncontentious. The Chinese counter-draft did not pursue its previous claim that all residents of Chinese heritage in British territories were still Chinese nationals to be governed in some degree by the Chinese authorities. The British government wisely avoided provoking the Chinese in this regard. It believed the matter could be appropriately handled by "a principle of international law that in the case of dual nationals the master nationality prevails, that is an Anglo-Argentinian is British in the United Kingdom and Argentinian in Argentina, and so with Anglo Chinese."[103] The British further wanted to insert an article to safeguard liberty of conscience and freedom of worship. The Chinese did not disagree with this. But they attempted to capitalise on this article so as to cover the right to establish schools and educate Chinese children on Chinese language and culture in British territories, principally those of Southeast Asia, in order to facilitate the inheritance of

---

**100** IMH Archives, 11–10–06–01–023, *Zhong-Ying shangyue* [The Sino-British Commercial Treaty], letter from the Ministry of Economy to the Ministry of Foreign Affairs, December 10, 1946.
**101** IMH Archives, 11–10–06–01–023, *Zhong-Ying shangyue* [The Sino-British Commercial Treaty], letter from the Ministry of Foreign Affairs to the Executive Yuan, December 19, 1946; IMH Archives, 11–10–06–01–022, *Zhong-Ying shangyue* [The Sino-British Commercial Treaty], letter from the OCAC to the Ministry of Foreign Affairs, November 7, 1944.
**102** IMH Archives, 11–10–06–01–023, *Zhong-Ying shangyue* [The Sino-British Commercial Treaty], letter from the Ministry of Foreign Affairs to Secretariat of the Executive Yuan, January 14, 1947.
**103** TNA, FO 371/63281, letter from A.L. Scott (FO) to W.A. Morris (CO), November 4, 1947.

Chinese identity (language, history and culture) by their descendants.[104] The British were prepared to accept the modification.[105]

In contrast to the rights of nationals, the treatment of companies raised much more controversy. First, in the view of the Chinese, the vast gap in development between Britain, one of the earliest industrialised nations, and China, a relatively backward agricultural country, meant that the treaty had to have some teeth to protect and nurture Chinese industries and public enterprises. Thus, the Chinese sought to avoid giving any concessions in this field to the British, even on the basis of reciprocity. Further, it was believed that Britain could only make a small contribution, through capital or technology, to Chinese future prosperity, particularly compared to the United States. It was therefore not worth conceding more favourable terms to the British, and also unwise, since the United States and other countries would enjoy the same treatment through the MFN clause.[106] The best terms the Chinese could offer were a form of limited national treatment for companies or MFN status, in conformity with the Sino-American treaty.[107] Yet, as far as the British were concerned, neither of these would provide "adequate" protection for their interests in China.[108]

In terms of taxation on nationals and companies, the Chinese, surprisingly, offered more favourable terms than the British could afford. Whereas London would only give MFN treatment for Chinese nationals (not companies) who were not residents in the United Kingdom, the Chinese formula offered national treatment to both nationals and companies and did not take into account where they were based. In addition, the Chinese draft allowed the British to enjoy national treatment in terms of domestic taxation of goods, whether imported or otherwise. However, the British had decided not to touch the provisions regarding goods despite the fact that this article was favourable to them.[109]

As regards non-profit institutions, the Chinese counter-draft did not mention the chambers of commerce or the trade guilds or associations which the British

---

**104** IMH Archives, 11–10–06–01–023, *Zhong-Ying shangyue* [The Sino-British Commercial Treaty], letter from the Ministry of Foreign Affairs to the Executive Yuan, December 19, 1946.
**105** TNA, FO 371/63281, "Note for consideration by an Inter-department Meeting to discuss further action on the Commercial Treaty with China", enclosed in the letter from H.O. Hooper (BT) to A.L. Scott (FO), October 11, 1947.
**106** IMH Archives, 11–10–06–01–023, *Zhong-Ying shangyue* [The Sino-British Commercial Treaty], letter from the Ministry of Economy to the Ministry of Foreign Affairs, December 10, 1946.
**107** IMH Archives, 11–10–06–01–023, *Zhong-Ying shangyue* [The Sino-British Commercial Treaty], letter from the Ministry of Foreign Affairs to the Executive Yuan, December 19, 1946.
**108** The British wanted full national treatment, as well as MFN status, in the area. See TNA, FO 371/63279, "Summary of main points" by the Board of Trade, February 20, 1947.
**109** Ibid.

had strived to protect through the treaty. But further deliberation made the British less insistent in this matter. John C. Hutchison, the British Minister (Commercial) in Shanghai, observed that the activities of the Chinese chambers of commerce in Malaya were mainly political in character. He was also informed by a Dutch acquaintance that the Chinese chambers of commerce in the Dutch East Indies were in fact controlled by the Overseas Affairs Department of the Guomindang. Hutchison accordingly questioned whether Article 16 (3) of the British draft – that "neither High Contracting Party will in his territories place any restrictions on the formation of chambers of commerce or similar un-incorporated bodies by the nationals of the other High Contracting Party" – was advantageous to the British.[110] Leo H. Lamb, a senior official in the British Embassy, also doubted whether the value of British chambers of commerce in China was sufficient to compensate for the unchecked political activities of Chinese chambers in Malaya and elsewhere.[111]

## 4.2 The Treatment of Goods, Copyright and Patent

The Chinese government was dissatisfied with the omission of articles relating to goods in the British draft, which in the former's eyes rendered the treaty incomplete.[112] Two articles were accordingly added to the treaty by the Chinese, specifying MFN treatment for import or export tariffs, and for import or export restrictions. However, the British refused to discuss this question because of their predicament regarding the balance of payment and worries that the matter may be affected by pending multilateral negotiations on trade policy.[113]

Like any developed nation, Britain wanted to safeguard its intellectual property, such as patents for inventions, trademarks, trade names, industrial designs and copyright for literary and artistic works. But for China, which desired to catch up with more advanced countries, this would greatly increase the cost of its learning. Thus, a reference to "industrial designs" was deleted from the Chinese counter-draft. In addition to this, there was controversy over the concept of translation. As far as the Chinese Ministries of Education and Justice were concerned, the freedom to translate was crucial for cultural intercourse and benefi-

---

110 TNA, FO 371/63281, telegram from J.C. Hutchison (Shanghai) to Leo H. Lamb (Nanjing), October 23, 1947.
111 TNA, FO 371/69625, telegram from L.H. Lamb (Nanjing) to A.L. Scott (FO), March 5, 1948.
112 IMH Archives, 11–10–06–01–022, *Zhong-Ying shangyue* [The Sino-British Commercial Treaty], minutes of the first conference of examining the British treaty draft, November 11, 1946.
113 TNA, FO 371/63279, "Summary of main points" by the Board of Trade, February 20, 1947.

cial to both countries. There was no legal or traditional prohibition in translating others' work in China. Chinese authors regarded the translation of their work by others as an honour and no one took account of the economic losses caused by any infringement of their copyright.[114] The Propaganda Department of the Guomindang frankly argued that freedom to translate would contribute to the Chinese absorption of "the fruits of Western academic work" and thus benefit from it.[115] The Chinese Ministry of Education suggested that the matter be left to the United Nations Educational, Scientific and Cultural Organisation (UNESCO) as both China and Britain were members of it.[116] However, this Chinese attitude towards translation was deemed ridiculous in British eyes: it seemed that "they [the Chinese] have the right to pirate copyright so long as this is translated, and they do not propose to offer any protection under their law for this."[117]

## 4.3 Other Commercial Matters

### 4.3.1 Immovable Property
The freedom to acquire, hold and dispose of lands and buildings in each country was an additional area of friction. The British were willing to open up their own real estate market fully to Chinese investors so as to secure the same conditions in China. Through such a concession, the British would "own, acquire, possess, hold, lease, hire and occupy property, movable or immovable, freehold or leasehold" in the Chinese territories and "dispose freely by sale, lease, hire, exchange, gift, marriage, testament, or in any other manner, of property, movable or immovable, lawfully possessed by them, or acquire such property by inheri-

---

114 IMH Archives, 11–10–06–01–022, *Zhong-Ying shangyue* [The Sino-British Commercial Treaty], letter from the Ministry of Education to the Ministry of Foreign Affairs, December 29, 1946; IMH Archives, 11–10–06–01–023, *Zhong-Ying shangyue* [The Sino-British Commercial Treaty], letter from the Ministry of Justice to the Ministry of Foreign Affairs, December 9, 1946.
115 IMH Archives, 11–10–06–01–022, *Zhong-Ying shangyue* [The Sino-British Commercial Treaty], letter from the Propaganda Department of the Central Executive Committee of the Guomindang to the Ministry of Foreign Affairs, December 3, 1946.
116 IMH Archives, 11–10–06–01–022, *Zhong-Ying shangyue* [The Sino-British Commercial Treaty], letter from the Ministry of Education to the Ministry of Foreign Affairs, December 29, 1946.
117 TNA, FO371/63279, "Summary of main points" by the Board of Trade, February 20, 1947.

tance."[118] However, the Chinese government did not wish to see foreigners acquiring more immovable properties in its territories, particularly land, although it had grudgingly agreed to this in the 1943 treaty. In fact, some subsequent domestic legislations had been introduced to regulate foreigners' landownership in China. For example, through the Chinese Land Law of 1946 certain categories of land were not allowed to be privately owned, such as land within a specified distance of the coast or the banks of navigable watercourses, water sources and highways used for public communications. Also, certain types of land were prohibited from being transferred, encumbered, or leased, to a foreign national, such as agricultural land, forests, fisheries, pastures, game reserves, salt beds, mines, water sources, forts, garrison areas and land on territorial borders.[119] Moreover, the Sino-American treaty only permitted American nationals, corporations and associations to lease land in China. Thus, there was little room for the Chinese to meet British requirements over this issue. In the Chinese draft, only the treatment related to MFN status regarding property was to be granted to British nationals and companies, and this was to be subject to the conditions and requirements as prescribed by Chinese laws. Unsurprisingly, this was "unsatisfactory" to the British authorities who required full and unconditional national treatment for its companies in these matters.[120]

### 4.3.2 Mineral Exploitation, Banking and Insurance

In the matter of mineral exploration and exploitation, the British focus lay on petroleum rather than other mineral resources. But the Chinese government had no intention of sharing its oil resources with London. The Chinese counter-article followed the Sino-American treaty. Under MFN treatment, the Chinese

---

118 TNA, FO 371/46223, "Draft Treaty of Establishment and Navigation between the United Kingdom and China (November 1945)" annexed to the letter from J.R. Willis (BT) to A.L. Scott (FO), November 13, 1945.

119 IMH Archives, 11–10–06–01–022, *Zhong-Ying shangyue* [The Sino-British Commercial Treaty], letter from the Land Department to the Ministry of Foreign Affairs, December 3, 1946; TNA, FO 371/63281, "Land Law (Revised and Promulgated by the National Government on 29 April 1946)" (hereafter "Land Law"), the English language version was annexed to the telegram from Ralph S. Stevenson (Nanjing) to Ernest Bevin (FO), August 21, 1946.

120 The Chinese treaty draft, for example, granted British nationals the right to lease (but not acquire) land. TNA, FO 371/63279, "Summary of main points" by the Board of Trade, February 20, 1947.

government reserved the power to close its mineral resources to all foreign countries.[121]

As for banking and insurance, they were at the core of British financial interests in China. Yet the Chinese omitted both matters in their draft.[122] The Board of Trade wanted an explicit safeguard in the treaty to protect the British insurance industry in China.[123] For its part, the Treasury, anxious about the impending Chinese Banking Law which contained some clauses which discriminated against foreign banks, believed that Britain should "endeavour to have [banking] re-instated."[124]

### 4.3.3 Shipping and Navigation

The matter of shipping and navigation created further disagreement. On the premise of reciprocating the same rights to the Chinese, the British draft proposed full national treatment, as well as MFN treatment, over the right of access of British ships to all ports and places in China; the right of commerce and navigation of British ships in Chinese waters; and the right to station, load and unload in Chinese ports. However, the Chinese draft offered none of the above, except under a conditional MFN status. Under such terms, British vessels would be allowed access to ports, places and waters subject to their being open to foreign shipping.[125] The British government could not accept these Chinese terms. After consulting with the shipowners, the Ministry of Transport asked that three points should be pressed:

a. The right for the British flag to trade from non-Chinese ports to all ports on the coast of China and to ports on the lower Yangtse River up to and including Hankou;
b. The right for the British flag to trade between Chinese coast ports and from Chinese coast ports to the river ports (mentioned in paragraph a) at least for 25 years, which was about the life of a ship;

**121** IMH Archives, 11–10–06–01–022, *Zhong-Ying shangyue* [The Sino-British Commercial Treaty], letter from the Land Department to the Ministry of Foreign Affairs, December 3, 1946.
**122** The Chinese treaty draft regarded banking, shipping and insurance as no different to other commercial activities. See TNA, FO 371/63281, letter from A.L. Scott (FO) to Miss P.M. Wack (BT), July 9, 1947.
**123** TNA, FO 371/63279, "Summary of main points" by the Board of Trade, February 20, 1947.
**124** TNA, FO 371/63281, letter from M. Loughnane (the Treasury) to Miss P.M. Wack (BT), June 16, 1947.
**125** TNA, FO 371/63279, "Summary of main points" by the Board of Trade, February 20, 1947.

    c. The right, accepting the loss of British flag right to trade between Chinese river ports, to form Chinese companies with British control but under the Chinese flag which will, of course, enjoy national treatment in trade between river ports as well as coastal ports.[126]

But the Chinese were in no mood to compromise. Institutions such as the Shanghai Shipping Guild lobbied the Guomindang authorities on this matter,[127] and, in any event, a Chinese Shipping Company Law prevented foreign investors from controlling or forming shipping companies under the Chinese flag. This law required that the Chinese should hold 60 per cent of the capital of a Chinese shipping company, as distinct from the 40 per cent requirement of Chinese Company Law in relation to other types of companies.[128]

## 5 "No Treaty Better than a Bad Treaty"

The Nationalist regime was eager to commence formal discussions over the treaty with the British, but the British were in no hurry. Consideration of the Chinese counter draft was handicapped by a shortage of expertise at the Board of Trade, which at this time was preoccupied with the International Trade Organisation (ITO) discussions at Geneva and Havana, and other trade agreements more closely affecting the economic life of Britain.[129] But British procrastination can

---

**126** TNA, FO 371/63279, draft cable to the British Ambassador to China, annex to the letter from G.V. Hole (MT) to G.V. Kitson (FO), February 5, 1947.

**127** IMH Archives, 11–10–06–01–022, *Zhong-Ying shangyue* [The Sino-British Commercial Treaty], letter from Shanghai Shipping Industry Guild to the Ministry of Foreign Affairs, November 16, 1946; TNA, FO 371/63280, "Shipping Industry Acquaints Government with Difficulties", *Financial Daily News* (March 14, 1947), translated and annexed to the telegram from Sir R.S. Stevenson (Nanjing) to Foreign Secretary (London), April 14, 1947.

**128** TNA, FO 371/63279, telegram from G.V. Kitson (FO) to Sir Ralph Skrine Stevenson (Nanjing), February 14, 1947; TNA, FO 371/63280, "Notes by Ministry of Transport on Board of Trade note (dated February 20, 1947)", attached to letter from R.C. Money (MT) to the Secretary (BT), April 3, 1947. In 1948, coastwise and inland water routes in China had been virtually monopolised by two Chinese companies: the state-owned China Merchants Steam Navigation Company and the private Minsheng 民生 Steamship Company, although their shipping tonnage was insufficient to meet demands. See TNA, FO 371/69559, telegram from H.A. Graves (British Embassy at Washington) to Mr. W. Walton Butterworth (US State Department), February 10, 1948.

**129** TNA, FO 371/63280, letter from H.O. Hooper (BT) to G.V. Kitson (FO), April 30, 1947; TNA, FO 371/69625, note by Foreign Office, February 26, 1948. The creation of the International Trade Organisation (ITO) failed in the end but the negotiations concluded the General Agreement on Tariffs and Trade (GATT).

also be attributed to its anticipation that no good results would come out of any Sino-British negotiations.

The Board of Trade speculated about difficulties in coming to terms with the Chinese government. The Chinese counter-draft was, it was opined, "in most respects wholly inadequate" to meet the British objectives. The Nationalists had followed very closely the terms of the commercial accord that they recently concluded with the United States which was deemed "a totally inadequate document".[130] Nevertheless, it was eager for the Chinese to be dragged back to the British draft.[131] Large British commercial interests were at stake. It was estimated that direct British investment was still worth £170 million after the war. This was largely through shipping companies, mining companies, public utility companies and merchant firms in China.[132]

Experienced China hands were not convinced that the Nationalists would concede. Ralph Stevenson and John Hutchison in the British Embassy suggested offering "a combined draft for presentation, as a basis for further study, to the Chinese authorities who should thus be more readily disposed to pursue negotiations in an objective and reasonable spirit."[133] At the same time, the British mercantile communities in China voiced their opinions through Hutchison:

First, … in existing economic and political circumstances the longer we could delay the conclusion of the new treaty the better;

Second, … in our treaty we should endeavour to obtain simplicity of language free from legalistic verbiage and capable of unequivocal translation from English into Chinese and vice versa;

Third, … we should, as far as possible, lay down broad principles within which trade could be carried on.[134]

---

**130** TNA, FO 371/63281, "Commercial Treaty Negotiations with China", a departmental minute by Commercial Relations & Treaties Department (BT), enclosed in the letter from H.O. Hooper (BT) to A.L. Scott (FO), October 11, 1947.
**131** TNA, BT 11/3466, "Minutes of Meeting held on Tuesday, 21st October, at the Board of Trade, to discuss the commercial treaty with China", October 21, 1947.
**132** TNA, FO 371/63281, "Commercial Treaty Negotiations with China", a departmental minute by Commercial Relations & Treaties Department (BT), enclosed in the letter from H.O. Hooper (BT) to A.L. Scott (FO), October 11, 1947. WWII caused little more than 11 per cent of losses to the value of British business investments in China. See Osterhammel, "China," 661.
**133** TNA, FO 371/69625, telegram from Ralph S. Stevenson (Nanjing) to Foreign Secretary (London), March 5, 1948.
**134** TNA, FO 371/63280, telegram from J.C. Hutchison (Commercial Counsellor, Shanghai) to the Chancery (British Embassy, Nanjing), March 18, 1947.

Leo Lamb, the British Minister in Nanjing, shared some of these views:

> there is some force in the argument that there may be certain advantages to be derived from avoiding undue haste so as to give time for the present mania for the preservation of sovereign rights, particularly insofar as they concern shipping, to subside, thereby affording a better chance of obtaining concessions from the Chinese negotiators. The prospects of introducing into our treaty just now privileges not contained in the American treaty are, at least, not enhanced by the fact that the terms of the latter document are still fresh in the memory of the general public.[135]

Soon the idea prevailed in London that it would be better to delay. Yet the Board of Trade was reluctant to agree to this approach. It would not be easy to explain publicly why negotiations had broken down and British trading interests would be left without any treaty protection other than the 1943 treaty.[136] The views of British diplomats in China worked to dispel such worries. Hutchison was confident that if questions were asked there would be little difficulty in explaining the position, namely that the negotiations had stalled but had not "broken down". He also confirmed that "our trading interests in China ... would certainly prefer no treaty to a bad treaty."[137]

The deteriorating political and economic situation in China further justified a postponement. Internal splits might give rise to a future repudiation of any Nationalists-signed agreement with the British; or the authorities in Nanjing might negotiate simply for terms which would appeal to the chauvinistic elements of the population as a diplomatic victory over a once powerful and imperialistic aggressor.[138] Nonetheless, pressure from Parliament hindered British officials from openly delaying the negotiations. The Foreign Office was clear that "any public statement indicating the desirability of delay would be impolitic."[139] In this regard, Stevenson's suggestion that "slowness is in fact the more expedient policy in existing circumstances in China" was accepted.[140] As a result, the treaty was put aside by departmental officials before any serious discussions in the Cabinet.

---

**135** Leo Lamb's words were relayed by G.V. Kitson, see TNA, FO 371/63280, letter from G.V. Kitson (FO) to G.E. Mitchell (China Association), May 27, 1947.

**136** TNA, FO 371/63281, "Commercial Treaty Negotiations with China", a departmental minute by Commercial Relations & Treaties Department (BT), enclosed in the letter from H.O. Hooper (BT) to A.L. Scott (FO), October 11, 1947.

**137** TNA, FO 371/69625, "Commercial Treaty Negotiations with China" by J.C. Hutchison, Minister (Commercial), Shanghai, February 15, 1948.

**138** Ibid.

**139** TNA, FO 37163280, telegram from G.V. Kitson (FO) to G.E. Mitchell (China Association), June 9, 1947.

**140** TNA, BT 11/3466, letter from P.W. Scarlett (FO) to G. Wilson (Cabinet Offices), April 9, 1948.

Ironically, the lack of a commercial treaty with China brought some advantages to British interests there. British shipping companies, for example, were able to continue to make their own arrangements locally. British merchants did not sell their shipping property along the Yangtse river, as stipulated in the 1943 Sino-British treaty.[141] Elsewhere, the British were able to obtain shipping facilities as *a quid pro quo* for practical assistance to Chinese shipping companies.[142] This could entail providing help to the Chinese in the training of seamen for both seagoing and shore crews.[143]

In March 1948, the Chinese government urged the British government to take part in formal treaty discussions, but with no effect.[144] Unknown to the Chinese, the Sino-British commercial treaty had already faded away from the British government's agenda. In this sense, the treaty had never been seriously negotiated.

## Conclusion

The outcome of the failure to agree a Sino-British commercial treaty can be largely laid at the doorstep of the British. The Nationalists were keen to settle the treaty at an early date. However, the British government was from the beginning unenthusiastic about a commercial treaty with China. Its participation in the negotiations can largely be attributed to the initiative taken by China and to being drawn down the path of the United States.

The Second World War gave rise to a more fragmented British Empire, albeit one which still sought to protect its interests around the world. London's power was no longer easy to project across the globe. The independence movements that boomed during the war meant that India, Burma and Malaya were more reluctant to accommodate themselves to a Sino-British commercial agreement. This diminished British bargaining power and rendered the reciprocal terms in the treaty less attractive to the Chinese government. Meanwhile, equality and

---

**141** TNA, FO 371/63280, "Shipping Industry Acquaints Government with Difficulties", *Financial Daily News* (March 14, 1947), translated and annexed to the telegram from Sir R.S. Stevenson (Nanjing) to Foreign Secretary (London), April 14, 1947.

**142** TNA, FO 371/69625, "Commercial Treaty Negotiations with China" by J. C. Hutchison, Minister (Commercial), Shanghai, February 15, 1948.

**143** TNA, FO 371/63279, letter from G.V. Kitson (FO) to Sir Ralph Skrine Stevenson (Nanjing), February 14, 1947.

**144** IMH Archives, 11–10–06–01–023, *Zhong-Ying shangyue* [The Sino-British Commercial Treaty], letter from the Ministry of Foreign Affairs to Chinese Embassy to London, March 19, 1948.

reciprocity became the main principles underlying the British approach, but not that of the Chinese. The British endeavoured to win back privileges lost in the 1943 treaty by way of offering the same rights to China and believed their companies could continue to play an important part in the Chinese economy, provided that the Chinese political and legal environment did not discriminate against foreign nationals and corporations. However, this gave rise to tensions over, for example, immigration control between the Board of Trade on the one side, and the Home Office and the Ministry of Labour on the other. As a result of opposition within the British imperial world, and disagreements inside Whitehall, British preparations for a draft treaty were substantially delayed.

At the same time, the Second World War caused a surge of nationalism in China. The Chinese wanted more protection in terms of their infant industries, rather than just an equal and reciprocal treaty. This was explicitly embodied in Chinese public protests against the Sino-American treaty. Under such conditions, any special British claims would be deemed unacceptable by the Chinese. Accordingly, the policy of "no treaty better than a bad treaty" emerged on the British side. Yet the two countries had bargained together like equal parties and the Chinese had not been forced to compromise under British military threats, as had happened in the negotiations over the Nanjing Treaty of 1842 and the Tianjin Treaty of 1858. In this sense, the era of "unequal treaties" was already past, although the attempt to conclude a commercial treaty had failed.

# Chapter III
# The Question of Hong Kong in Sino-British Relations, 1942 – 1945

A number of academic works have touched on wartime Sino-British conflicts over Hong Kong, but these have often dealt with the issue within a wider context of imperial affairs during this era.[1] Relatively few works, for example, explore the debate amongst the Nationalists over raising the matter of the New Territories in the 1943 treaty negotiations, and the shift in the official Nationalist attitude towards Hong Kong during the war.[2] In addition, British clandestine plans for the purpose of reoccupying the colony ahead of Chinese forces have not been addressed in the existing scholarship; however, these provide vital evidence to demonstrate that the British success of reoccupying Hong Kong in August 1945 was unexpected, mainly owing to the suddenness of the Japanese capitulation and the support of Washington. This chapter endeavours to fill these lacunae as well as provide an overview of Sino-British disputes regarding the colony's future during the years 1943 – 1945 by referencing both Chinese and British official and private documents.

After a brief introductory section, the chapter addresses Sino-British discussions over the early return of the New Territories of Hong Kong, which were part of the 1943 Sino-British Treaty negotiations and have been referenced in passing in Chapter I. It was the first time that the Chinese had raised the issue of the sovereignty over Hong Kong with the British government since the colony had been ceded in the nineteenth century. However, these discussions ended with no result except for a British promise to discuss the matter after the war. Thereafter, Chongqing and London were plunged into competition for the sovereignty of

---

1 See, for example, Chan, *China, Britain and Hong Kong, 1895 – 1945*, 71 – 89; ibid., "Hong Kong in Sino-British Diplomacy, 1926 – 45."; Liu, *A Partnership for Disorder*, 118 – 139; Li, *Zhanshi Yingguo dui Hua zhengce [Britain's China Policy during the Second World War]*, 170 – 185; Whitfield, *Hong Kong and the Anglo-American Alliance*, 61 – 211; Tsang, *Hong Kong: An Appointment with China*, 34 – 53; ibid., *A Modern History of Hong Kong*, 121 – 138; Shai, *Britain and China, 1941 – 47*, 28 – 94.

2 Steven Tsang has somewhat discussed the latter question. After examining a memorandum, drafted in September 1944 by the European Department of the Chinese Foreign Ministry, and two documents revised and expanded from it by the same author in August 1945, he came to a conclusion that the Nationalist diplomatic system had no feasible plan to assume Chinese authority over either the New Territories or the whole of Hong Kong after the war. See Tsang, *Hong Kong: An Appointment with China*, 37.

https://doi.org/10.1515/9783110706659-010

Hong Kong. The chapter focuses on the Nationalist effort to enlist American support in returning Hong Kong to China, as well as British responses to this approach, while also addressing British secret plans for resuming direct British authority over the territory. Japan's sudden capitulation in August 1945 offered the British a chance to achieve this aim with the help of their naval forces. However, as the chapter reveals, a new argument with the Chinese then broke out over the matter of who had the right to accept the Japanese surrender in Hong Kong.

# 1 The British Colony of Hong Kong

Hong Kong was a patchwork comprised of three parts. Hong Kong Island and the Kowloon Peninsula (including Stonecutters Island) were ceded permanently to Britain through the Treaty of Nanjing in 1842 and the Treaty of Beijing in 1860 respectively. The Kowloon Leased Land or the New Territories was leased to Britain via the Beijing Convention of 1898. Compared to the two ceded areas, the New Territories were complex in legal terms since in the Sino-British agreement of 1898 full sovereignty over the land was not given to the British. First, the New Territories were leased to Britain, for the purpose of "the proper defence and protection of the Colony", as an extension of the British colony of Hong Kong for 99 years. Second, Britain's "sole jurisdiction" over the New Territories exempted the Kowloon Walled City, the seat of the Chinese government in the area. The Chinese authorities retained their jurisdiction within the city, the right to utilise the existing landing place near the city, the right to travel along the road between Kowloon and Xin'an County 新安縣 (named as Bao'an County 寶安縣 after 1914) and the right to use the waters of Dapengwan 大鵬灣 (Mirs Bay) and Shenzhenwan 深圳灣 (Deep Bay). Third, the convention was ambiguous in relation to British authority over expropriating lands and expelling inhabitants from the New Territories.[3]

Later developments further complicated the Kowloon Walled City's status. Although the leasehold was legitimised by the Chinese government, the British encountered vehement resistance from Chinese locals. As a response, the British authorities waged a series of military campaigns in 1899 and expelled Chinese officials and troops from the Walled City of Kowloon as well as Chinese customs houses. The city was believed to be a spiritual symbol, or even a support base,

---

3 UK Treaties Online, "Convention between the United Kingdom and China respecting an Extension of Hong Kong Territory," 1898, http://foto.archivalware.co.uk/data/Library2/pdf/1898-TS0016.pdf (accessed April 7, 2019).

**Map 2:** Map of Hong Kong.
© Zhaodong Wang & © Peter Palm, Berlin.

for local Chinese defiance. By claiming that a continual Chinese presence in the Kowloon Walled City was inconsistent with the military requirements of the defence of Hong Kong, the Hong Kong council directed that "the City of Kowloon shall be ... part and parcel of Her Majesty's Colony of Hong Kong."[4] However, no Chinese government ever acknowledged this edict.

The Chinese government had never formally raised the restoration of Hong Kong with Britain before 1942.[5] Sino-British disputes over Hong Kong had occurred in the 1930s but were restricted to the sovereignty of the Kowloon Walled City. Nevertheless, the return of Hong Kong to China was a Nationalist aim. As early as January 1925, Sun Yat-sen, the founder of the Guomindang, had disclosed this intention in private talks with Tōyama Mitsuru, a Japanese activist, in Kobe.[6] Chiang Kai-shek, Sun's successor, held the same view. In his book, *China's Destiny*, he remarked that

> What we must recognise is that there exists a definite geographical interdependence between Hong Kong [the ceded Hong Kong] and Kowloon [the New Territories], and that both questions must be settled at the same time.[7]

The Nationalists, particularly Chiang Kai-shek, believed that the Pacific War and the treaty abolishing British extraterritoriality in China offered an opportunity to fulfil this ambition.

---

4 "The Walled City Order in Council (27 December 1899)", see Appendix 3 in Wesley-Smith, *Unequal Treaty 1898–1997*, 196–197.

5 James T. H. Tang, "World War to Cold War: Hong Kong's Future and Anglo-Chinese Interactions, 1941–55," in *Precarious Balance: Hong Kong between China and Britain, 1842–1992*, ed. Ming K. Chan (New York and London: M.E. Sharpe, 1994), 109; Kevin P. Lane, *Sovereignty and the Status Quo: the Historical Roots of China's Hong Kong Policy* (Boulder, San Francisco and Oxford: Westview Press, 1990), 42–43; Tsang, *Hong Kong: An Appointment with China*, 25.

6 Gengxiong Wang, ed., *Sun Zongshan ji wai ji [The Supplementary Anthology of Dr. Sun Yat-sen]* (Shanghai: Shanghai Renmin Chubanshe, 1990), 318. The genuine motive behind Sun referring to Hong Kong was not clear. Perhaps he also attempted to utilise the existing antipathy between Britain and Japan in East Asia to solicit the latter's military or financial support.

7 Kai-shek Chiang, *China's Destiny and Chinese Economic Theory* (New York: Roy Publishers, 1947), 154. The original Chinese-language version was first published in 1943.

## 2 Disputes over the New Territories in the 1943 Sino-British Treaty Negotiations

As we have seen, Britain and China had a fundamental divergence of opinion over the 1943 treaty. Britain was forced to follow in the Americans' footsteps and its intention was to give away as few treaty rights in China as possible. In contrast, the Chinese hoped for a comprehensive sweeping away of the existing unequal treaties. The restoration of Hong Kong was an indispensable part of this agenda. Chiang also wanted to include the status of Tibet in the negotiations but was dissuaded by his officials on the grounds that Tibet was more related to the question of the Chinese frontier than extraterritoriality.[8]

In relation to Hong Kong, the Nationalists chose to raise only the issue of the return of the New Territories (also known as the Kowloon Leased Land) rather than the other parts of Hong Kong. It was believed that the New Territories were in the same category as concessions and settlements, since they were all leased to the British. Further, the New Territories, comprising 80% of Hong Kong's territory and holding the main sources of fresh water and the sole airfield in the colony, were deemed as a good starting point for the retrocession of the whole of Hong Kong to China. Nevertheless, there were some differing views inside the Chinese government.

### 2.1 Nationalist Opinions on Raising of the Issue of the New Territories

The Nationalists' thinking about their relations with Britain was mainly dominated by two schools. The first was headed by Gu Weijun, who had represented China at the Paris Peace Conference in 1919 and was the Chinese Ambassador to London during the Second World War. Gu endorsed a close Sino-British collaboration and, as a senior professional diplomat since the start of the Republic and seldom involved in power politics within the Guomindang, he had some influence over Chiang. During the negotiations, Gu was at Chongqing accompanying a British parliamentary delegation visiting China, and fully participated in the Sino-British discussions. In Gu's view, the establishment of a Sino-British-American alliance was key for China in order to maintain an advantageous status in the post-war world because such alliances would inevitably dominate global politics when peace was restored. He admitted that the Chinese would

---

**8** October 31, 1942, see Lin, ed., *The Diary of Dr. Wang Shih-chieh*, 465; Koo, *Gu Weijue huiyilu vol. 5*, 101.

gain little from a Sino-British alliance during wartime but, without an association with Britain, China would be in danger in the post-war era. As he observed, there were plenty of pro-Japanese officials working in Whitehall who would easily establish a close relationship with post-war Japan to contain China. Only if Chongqing maintained good terms with London during the war could it prevent such Anglo-Japanese connections. More importantly, China could not ally with the United States and meanwhile turn its back on Britain. If the Americans had to make a choice in a Sino-British struggle, they would definitely choose Britain.[9]

Wang Shijie, the Secretary-General of the Central Planning Board and later Chinese Foreign Minister, was broadly in line with Gu. But he was concerned more about the danger from the Soviet Union. In his estimation, Sino-Soviet conflicts would be a threat during the post-war era, since there were a large number of thorny questions between China and the Soviet Union, such as Outer Mongolia, the Chinese Communist Party and Northeast China including Lüshungang 旅順港 (Port Arthur) and Dalian. It would thus be unwise to offend Britain and lose a potentially sympathetic supporter in any subsequent peace negotiations.[10]

Song Ziwen, Chiang's brother-in-law and the then Chinese Foreign Minister, led the other school. He doubted that the White House had an unbreakable partnership with Downing Street. As far as China was concerned, he saw few advantages in a close Sino-British relationship and believed that it would undermine China's image among other Asian countries. In Song's view, the era of the British Empire had passed, and the maintenance of a close association with the United States was crucial for China.[11] Indeed, Song's anti-British sentiment chimed with Chiang's impression that Britain was a stubborn and arrogant colonial country. This was amplified by the latter's trip to India and the difficult Chinese cooperation with British troops in Burma.[12] These kinds of emotions pervade Chiang's talks, diaries and his book, *China's Destiny*.[13]

Some of these tensions were hinted at in internal discussions about the New Territories. In October 1942, high-level Chinese officials convened at Chiang's residence in Chongqing to discuss the British treaty draft. The majority agreed with a proposal to seek to restore the Kowloon leased area, rather than the entirety of Hong Kong, to China. But Fu Bingchang 傅秉常 (Foo Ping-sheung), the Vice-Minister for Foreign Affairs, showed some hesitation due to the difficulties

**9** Ibid., *Gu Weijue huiyilu vol. 5*, 98–100.
**10** May 12, 1942, see Lin, ed., *The Diary of Dr. Wang Shih-chieh*, 430.
**11** Koo, *Gu Weijue huiyilu vol. 5*, 98–99.
**12** Chan, "The Abrogation of British Extraterritoriality," 260–263.
**13** Chiang, *China's Destiny and Chinese Economic Theory*, 44–107.

of connecting the Kowloon leasehold with the subject matter of the treaty: extra-territoriality.[14] This was a position shared by Gu.[15] In contrast, Wang Shijie saw a link. He regarded the leasehold of Hong Kong as similar to Qingdao 青島 (Tsing-tao) and Weihaiwei 威海衛, since these lands were all leased for military purpos-es and had already been returned to China.[16] However, Gu could not agree with Wang. In his view, the core difference between Weihaiwei and the New Territo-ries was not about military use but lease tenure.[17] According to the convention, Weihaiwei was leased to Britain as a naval base for as long a period as Russia occupied Lüshungang.[18] After Japan defeated Russia and took over Lüshungang in 1904–05, there had been no legitimacy for the British to keep the territory al-though it did not return to China until 1930.

## 2.2 The Legal Status of the New Territories

The Nationalist tactic of linking the matter of the New Territories to extraterritor-iality derived from a hasty decision. In legal terms, the two matters had little con-nection. The New Territories were leased to Britain as a portion of its territory for 99 years for the purpose of "the proper defence and protection of the Colony".[19] Although it was debatable whether the British government had full sovereignty over these lands, the convention had explicitly distinguished the New Territories from those concessions or settlements which were leased by foreign nations ac-cording to the stipulations of the treaty-port system and governed by foreigners on the basis of extraterritorial rights.

To a large extent, as Wang Shijie argued, the New Territories had a similar status to that of Weihaiwei. It was with a military intention that Britain had rent-ed Weihaiwei, a seaport in North China, as a naval base in 1898. Furthermore, the British had sole jurisdiction over this area with the exception of the walled

---

**14** October 31, 1942, see Lin, ed., *The Diary of Dr. Wang Shih-chieh*, 465.

**15** Koo, *Gu Weijue huiyilu vol. 5*, 100–101.

**16** October 31, 1942, see Lin, ed., *The Diary of Dr. Wang Shih-chieh*, 465. Qingdao or Jiaozhouwan 膠州灣 was leased in 1898 to Germany as its concession and maritime base for 99 years. During the First World War, the city was occupied by the Japanese and did not revert to China until the Washington Conference (1921–22).

**17** Koo, *Gu Weijue huiyilu vol. 5*, 163.

**18** UK Treaties Online, "Convention between the United Kingdom and China respecting WEI HAI WEI," 1898, foto.archivalware.co.uk/data/Library2/pdf/1898-TS0014.pdf (accessed May 29, 2019).

**19** Ibid., "Convention between the United Kingdom and China respecting an Extension of Hong Kong Territory".

city of Weihaiwei. Nevertheless, the vital difference was not use, or tenure as Gu had pointed out, but the additional wording in the convention of the New Territories, which stated that "the limits of British territory shall be enlarged under lease."[20] On the basis of this phrase, Whitehall regarded the leased land as an integral part of the colony of Hong Kong during the term of lease. Moreover, in terms of the style of local British administration, it could be argued that there was a distinction between Weihaiwei and the New Territories. The Colonial Office had intended to assume and maintain full sovereign rights over Weihaiwei but had drawn back from this due to fears that Germany, Russia and France would follow this precedent. Thus, Weihaiwei was treated as foreign soil and was not officially recognised as part of the British domains. In contrast, due to the nature of the territory's acquisition by military force, the New Territories were treated as part of the colony of Hong Kong although in practice the leased land was administered as a separate area.[21]

Therefore, according to the analysis above, it would have been hard for the Chinese to engage with the Kowloon question on the basis of similarities with Weihaiwei or extraterritoriality. The rational argument for the Chinese would be to treat the New Territories as a separate case. On the basis that they were no longer meaningful for the defence of the Hong Kong colony owing to the advance of military technology and the alliance between China and Britain, Chongqing could ask for a shortening of the tenancy. Or, as the Communists government would do in the 1980s, it could deem all treaties concerned with Hong Kong unequal and thus invalid. However, poor preparations had placed the Nationalist demand for the New Territories on a weak footing before the negotiations began.[22]

## 2.3 Sino-British Discussions over the New Territories in 1942

In a counter-draft to the British one in November 1942, the Chinese government raised the issue of the termination of the 1898 Sino-British convention. Although it had anticipated such a move, Whitehall hesitated about how to react. After a debate among officials in London, the government not only insisted that the New Territories had "nothing to do with extraterritoriality" but also being "an enlargement of British territory is in an entirely different category from the conces-

---

**20** Ibid.
**21** Wesley-Smith, *Unequal Treaty 1898–1997*, 90–92.
**22** On the question as to why the 1898 convention was an unequal treaty, see Wesley-Smith, *Unequal Treaty*, 184–187.

sions and settlements in China."[23] The Foreign Office suggested a face-saving compromise: the Chinese government could initiate discussions about the future status of the New Territories after the war, but without reference to the termination of the lease.[24] However, Song Ziwen and his colleagues, former and current Vice-Ministers, Fu Bingchang and Wu Guozhen 吳國楨 (K. C. Wu,),[25] were stubborn on the matter:

> the Chinese public regarded leased territories as in the same category as concessions, that the matter had been raised in the People's Political Council, that it was desirable to remove all causes of misunderstanding between the two peoples, and that the Chinese Government felt that a treaty which did not secure a settlement of [the] Kowloon lease question would fail to achieve this object.[26]

Yet there seemed to be some room for flexibility behind the scenes. Song confided to the British ambassador via Hang Liwu 杭立武, a professor at Central University of China who had close links with the British government, that he had little confidence in convincing Chiang Kai-shek and Kong Xiangxi 孔祥熙 (Kung Hsiang-hsi), Chiang's influential brother-in-law and then the Chinese Minister of Finance, to conclude the treaty without the inclusion of the Kowloon leasehold. Hang suggested a middle course whereby "the Chinese Government should address to His Majesty's Government a communication stating that, while recognising that the question was not concerned with the matter now under negotiation, they desired to raise it later on at a more appropriate time."[27] Whether this solution had Song's formal endorsement was not clear, and it was unlikely that Chiang Kai-shek would accept such a formula. Nonetheless, Anthony Eden and Horace Seymour were encouraged. In their minds, a proposal from a Chinese scholar with close connections with the top-level would surely be acceptable to the Chongqing government. Influenced by this conviction, Eden raised the proposal with the War Cabinet and the latter accepted it.[28]

---

23 TNA, CO 129/588/24, telegram from Foreign Office to Chongqing, December 4, 1942.
24 Ibid; Anthony Best, ed., *British Documents on Foreign Affairs – Reports and Papers from the Foreign Office Confidential Print, Part III (1940–1945), Series E (Asia)*, 8 vols., vol. 6 (Bethesda, MD: University Publications of America, 1997), no. 28, 76 (hereafter *BDFA, Part III, Series E, vol. 6*).
25 Fu Bingchang left the post of Vice-Minister for Foreign Affairs on December 8, 1942 and was appointed as the new ambassador to Moscow. Wu Guozhen succeeded him in the post.
26 Wu's words were relayed by Seymour. See TNA, CO 129/588/24, telegram from Sir H. Seymour to Foreign Office, December 15, 1942; *BDFA, Part III, Series E, vol. 6*, no. 36, 88.
27 Ibid.
28 According to the Cabinet minutes, Churchill emphasised that there would be no consideration of the question of territorial adjustments during the war and the Kowloon issue must be left

In December 1942 London made its position clear to the Nationalists. There were two main points: "the future of the New Territories is outside the scope of the present agreement"; and "if the Chinese Government desires that terms of the lease of these territories should be reconsidered, that is a matter which ... should be discussed at the Peace Conference."[29] The wording "at the Peace Conference" was subsequently replaced with the more ambiguous phrase of "when victory is won". In discussion, Song was inclined to accept this proposal, although he and his colleagues were suspicious of the subtle meaning of "terms of lease", which could imply a continuance of the lease subject to unspecified modifications.[30] In order to meet these Chinese misgivings, the British modified the words "terms of lease" to "lease" or "question of lease".[31]

But the Chinese response was ultimately in the hands of Chiang Kai-shek.[32] A weak stance on his part would disappoint Chinese nationals and embarrass his government as the guardian of the national interest, especially if Wang Jingwei's 汪精衛 puppet regime was able to abrogate more foreign privileges with the aid of Japan. Yet there were worries about pressing the issue too far at this point. In talks with Song, Chiang insisted if there was to be a compromise, the question should be dealt with in an exchange of notes affixed to the treaty rather than as the separate exchange of letters that the British wanted.[33] And London should clearly show its willingness to return the New Territories and indicate it was ready to discuss this at any time. Indeed, in his diaries, Chiang chided the British for "deliberately making the negotiations difficult", but, interestingly, he was optimistic as to the outcome "because the Americans still had their sincerity."[34]

Gu formally conveyed Chiang's thinking to Seymour. There would be no Chinese objection to seeing the matter dealt with outside the treaty and settled after the war. But Whitehall should make clear its intention to return the New Terri-

---

to a peace conference. See TNA, CO 129/588/24, "extract from conclusions of a meeting of the War Cabinet held at 10 Downing Street, S. W. L.", December 21, 1942.

**29** Telegram from Mr. Eden to Sir H. Seymour, December 23, 1942, Best, ed., *BDFA, Part III, Series E, vol.* 6, no. 46, 94.

**30** TNA, CO 129/588/24, telegram from Sir H. Seymour to Foreign Office, December 25, 1942; ibid., no. 49, 95.

**31** TNA, CO 129/588/24, telegram from Foreign Office to Chongqing, December 28, 1942; ibid., no. 57, 99.

**32** During the war, Chiang headed the central government at Chongqing, even though there were some local dissident Nationalist commanders. In terms of diplomatic affairs, final decision-making was in his hands, but he frequently conducted private meetings with his most trusted diplomats and other advisors.

**33** December 26, 1942, see Gao, ed., *SLGB vol. 52*, 119–121.

**34** Ibid.

tories to China in writing. Otherwise, the Chinese government would not sign the treaty.[35] The British, however, were not going to back down over this. Eden instructed Seymour that "the solution proposed by Dr Koo [Gu] is unacceptable to us, and if [the] Chinese persist we shall have to do without the Treaty."[36] Under the circumstances, the Sino-British negotiations appeared to be on the point of collapse.

Neither the Chinese nor British governments relished the prospect of a broken relationship, but this did not mean that they believed Sino-British relations were of critical importance to their respective national interests. To the Chinese, Britain was contributing little to their military resistance against the Japanese and was even preventing more American resources being invested in the China theatre. To British eyes, the Nationalist government was not a much lesser threat to the British Empire in East Asia than Japan. But both governments regarded their alliance with America as a top priority. As Wang Shijie pointed out, Chongqing could not maintain a close partnership with Washington while breaking up its alliance with London.[37] This perhaps explains why Chiang was willing to compromise because he worried about the attitude of the Americans if a Sino-British treaty fell apart due to his intransigence over the New Territories, which were still under Japanese control. It also reflected the fact that Chiang did not have full confidence in enlisting American support over the matter of Hong Kong at this time.

A number of scholarly works indicate that the Chinese suddenly and surprisingly acquiesced to the British position on December 31, 1942.[38] In fact, Chiang's decision to compromise was taken on December 27, before the British sought the support of the US. On the evening of that day, after having dinner with Song Ziwen and Gu Weijun, Chiang finally gave up his claim to the New Territories. In his diaries, he remarked:

> It would indeed give the British a blow if the Sino-British treaty could not be signed simultaneously with the Sino-American one. But the treaty is of importance to China. In compar-

---

35 TNA, CO 129/588/24, telegram from Chongqing to Foreign Office, December 27, 1942.

36 TNA, CO 129/588/24, telegram from Foreign Office to Chongqing, December 28, 1942; Best, ed., *BDFA, Part III, Series E, vol. 6*, no. 57, 99.

37 May 12, 1942, see Lin, ed., *The Diary of Dr. Wang Shih-chieh*, 430.

38 See Chan, "The Abrogation of British Extraterritoriality," 289; Lane, *Sovereignty and the Status Quo*, 48; Tsang, *Hong Kong: An Appointment with China*, 32. Li Shi'an otherwise advanced the date to December 25, 1942. See Li, "The Extraterritoriality Negotiations of 1943 and the New Territories," 637–638.

ison with this, the Kowloon question is a partial problem. I should consider it against a broader background and avoid doing harm to the Alliance.[39]

Whether there was more to Chiang's thinking here remains unclear. Chiang himself once put the blame on *Zhongyang Ribao*, an official newspaper of the Chinese Nationalist Party. Without authorisation, the paper published an article on December 27 claiming that the Chinese government had agreed to sign treaties with both the Americans and the British on January 1, 1943, although at the time the Sino-British negotiations were still stalled.[40] Yet this public pronouncement, though undoubtedly serious, seems insufficient to have swayed Chiang, a nationalist with a strong anti-British streak. A more convincing explanation might lie in the potential loss of a Sino-British treaty which was in general terms advantageous to Chinese interests.[41]

In spite of Chiang's decision, Song and his colleagues did not want to give way to the British immediately.[42] Since there were still a few days left until January 1, Song intended to push Seymour further in the hope of a more advantageous agreement. He demanded that Britain "should agree to declare now their readiness to return leased territories to China", within or separate from the treaty, and "any settlement which left out [the] Kowloon lease would fail in its object of establishing [Sino-British] relations on a basis of mutual confidence." Unsurprisingly, Seymour rebuffed this formula, but he was uneasy about a possible failure of the treaty. As far as he was concerned, such an outcome would be a blow to Britain, especially if "the Chinese would, after securing [the] American treaty, proceed to the unilateral abolition of extraterritorial rights and even if they did not do this, the position would be very difficult and lead constantly to friction."[43]

Back in London, Eden urgently instructed Viscount Halifax, the British ambassador to Washington, to enlist the United States' help in order to put pressure on the Chinese. It was conveyed that Britain had endeavoured to coordinate its actions with the United States, but its policy over retaining the New Territories was not to be challenged.[44] The Americans had no similar areas of dispute

---

**39** December 27, 1942, see Gao, ed., *SLGB vol. 52*, 123.
**40** Koo, *Gu Weijue huiyilu vol. 5*, 168.
**41** Ibid., 164.
**42** Ibid., 163.
**43** TNA, CO 129/588/22, telegram from Sir H. Seymour to Foreign Office, December 28, 1942; Best, ed., *BDFA, Part III, Series E, vol. 6*, no. 56, 98.
**44** TNA, CO 129/588/24, telegram from Foreign Office to Washington, December 29, 1942; ibid., no. 60, 110–111.

with China, but they were displeased with the Chinese for frequently introducing extraneous issues into the treaty discussion. The US State Department was therefore prepared to inform the Chinese government in relation to the New Territories question that: "it was of concern to the United States Government in so far as its introduction might impair the smooth settlement of extraterritorial question, upon which the policy of the three countries was in harmony." Meanwhile, the United States decided to postpone the signing date from January 1 to January 5, 1943 "owing to difficulties in regard to collation of English and Chinese texts", leaving more time for Britain and China to resolve their disputes over Hong Kong.[45]

The US disapproval of introducing the matter of the New Territories into the treaty changed the dynamics of the debate. Seymour turned to insist that the Chinese government, rather than the British government, clarify that the New Territories were not connected to extraterritoriality and that it was the Chinese desire "to raise the question at ā more appropriate time".[46] This made Chiang furious:

> [The British government] did not only disagree with [dealing with Hong Kong by] the means of an exchange of notes, but also dispensed with their previous text and even required the Chinese government to make a statement that the Kowloon question was outside of the unequal-treaty matters and not in the scope of the treaty negotiations. If this could be borne, what else could not be borne! Since I was all ready to sign the treaty as a main goal, I would not argue anymore over this. In the future, I will make a statement [to reserve the rights related to the New Territories]. Gu Weijun had advised me of making the statement before the signing, as a base for later negotiations. However, in my opinion, there was no such need. After the formal signing, Hong Kong must be occupied by our troops first. At that time, it would not be an issue whether we had made such words.[47]

However, due to the approaching deadline, Chiang accepted the terms. Song immediately informed Seymour of the Chinese government's decision.[48] The Chinese were now eager to meet the January 1, 1943 deadline.[49] When Song discovered that the Americans had deferred the signing date he was angry and chided

---

**45** TNA, CO 129/588/24, telegram from Viscount Halifax (Washington) to Foreign Office, December 31, 1942; Koo, *Gu Weijue huiyilu vol. 5*, 167–168.

**46** Two telegrams from Sir H. Seymour to Mr Eden, December 30, 1942, see Best, ed., *BDFA, Part III, Series E, vol. 6*, no. 63 and 64, 112–113.

**47** December 30, 1942, see Gao, ed., *SLGB vol. 52*, 138.

**48** Telegram from Sir H. Seymour to Mr Eden, December 31, 1942, see Best, ed., *BDFA, Part III, Series E, vol. 6*, no. 65, 113.

**49** The Chinese government agreed that there would be no exchange of notes on the subject of the New Territories, but it reserved the right to raise the question later. See Telegram from Mr Eden to Viscount Halifax (Washington), December 31, 1942, ibid., no. 66, 113.

the Chinese Ambassador to Washington Wei Daoming 魏道明 for not informing him promptly.[50]

The treaty negotiations were now over. Although many aspects of this treaty were highly satisfactory to the Chinese, the Hong Kong issue clearly rankled with Chiang. Yet he predicted that the military events might help him to restore this territory:

> After waking up at 5 am [December 31], I started to consider the Hong Kong issue. I do not require the British government to make any promise about the Kowloon matter [the New Territories]. On the contrary, the British are pressuring me into declaring that Kowloon [leased territories] does not belong to the unequal treaties system; otherwise they will not sign the new treaty. If that is the case, the only thing that our government should do is to issue a statement on the abolition of unequal treaties and the negation of British historical rights in China. As long as we occupy Hong Kong with military force from the Japanese hands, the Britons, although cunning, would certainly be at a loss. This is a last resort.[51]

In January 1943, one day after the signing of the treaty, the British ambassador received an official Chinese note (dated January 11), stating that the Chinese government reserved its right to discuss the future of Hong Kong's leased territories "at a later date".[52] At the same time, Song Ziwen explained publicly that "the question of Kowloon leased territories had been raised by China but at [British] instance not discussed and China reserved the right to revive the question later." These initiatives, however, did not receive any response from London.[53] In Anthony Eden's eyes, silence appeared to be the best way to keep Hong Kong off the international agenda.

## 3 The Mediation of the United States over the Matter of Hong Kong, 1943–1945

After the treaty negotiations, the Nationalist government did not abandon its efforts to regain Hong Kong. Two factors were in its favour. First, American anticolonial sentiment reached a wartime peak in 1942–1943.[54] The US suspected

---

**50** Koo, *Gu Weijue huiyilu vol. 5*, 167–168.
**51** December 31, 1942, see Gao, ed., *SLGB vol. 52*, 140.
**52** TNA, FO 371/35680, telegram from Dominion Office to Canada, Australia, New Zealand and South Africa, January 15, 1943.
**53** Ibid.
**54** Louis, *Imperialism at Bay*, 11.

Britain of "conducting its war effort mainly in the interest of the empire."[55] This led to a fear on the part of the British that the United States would liberate British colonies and not return them to the former's rule.[56] Second, Hong Kong lay in the Allied China Command theatre, the supreme commander of which was Chiang Kai-shek.[57] He thus had some reason to believe that Chinese troops might liberate the city from the Japanese occupation. If that was the case, the Nationalist would undoubtedly have an upper hand in any ensuing negotiations with the British. In the meantime, the main Chinese approach was to solicit US intervention.

When Madame Chiang undertook a visit to Washington in early 1943, Chiang asked her informally to test President Franklin Roosevelt's attitude towards the future of Hong Kong. After confirming his generally positive response, Chiang instructed Song Ziwen to raise the issue formally with the president.[58] Roosevelt's plan to resolve the issue was that if Britain was willing to return all of Hong Kong to China the latter should declare Hong Kong a free port (free to the commerce of all nations) and promise to protect the interests of expatriate Britons. Chiang asked his Supreme National Defence Council of the Guomindang (SNDCG) to consider the proposal. In March 1943, the plan was agreed.[59] Meanwhile, Roosevelt discussed the proposal with Anthony Eden when the latter was paying an official visit to Washington, but without any agreement.[60] Despite this, Roosevelt and his top-level officials, exemplified by Sumner Welles, the Under-Secretary of State, were confident that Britain would finally come in line with this idea if the Chinese made more concessions.[61]

---

**55** Schake, *Safe Passage*, 266.

**56** Ibid., 267.

**57** The China Theatre at first consisted of China, Indo-China and Siam (Thailand). After the British South East Asia Command (SEAC) was established in 1943, Siam was separated from the China Theatre and incorporated into SEAC, as well as part of Indo-China (south of the sixteenth parallel) in August 1945.

**58** March 13, 1943, see Sulan Gao, ed., *Jiang Zhongzheng zongtong dang'an: shilüe gaoben [Draft Papers of Chiang Kai-shek]*, 82 vols., vol. 53 (Taipei: Academia Historica, 2011), 28; Tsang, *Hong Kong: An Appointment with China*, 35.

**59** The Chinese government agreed to declare Hong Kong (both leased and ceded territories) as a free port after Britain had returned the colony, noticeably on its own initiative rather than as an *a priori* condition for the British retrocession. See AH Archives, 001–064520–0002, *Zujie shouhui* [The Recovery of the Leased Territories], letter from the SNDCG to the Nationalist government, March 20, 1943; March 20, 1943, see Lin, ed., *The Diary of Dr. Wang Shih-chieh*, 495.

**60** Lane, *Sovereignty and the Status Quo*, 49. However, Xiaoyuan Liu argues that Roosevelt did not discuss the matter with Eden. See Liu, *A Partnership for Disorder*, 138.

**61** Koo, *Gu Weijue huiyilu vol. 5*, 326–327.

In July-August 1943, Song Ziwen did not raise the matter of Hong Kong when he visited London, which surprised the British. Clearly, he wished to avoid further disturbing the already-strained Sino-British relations. His main aim on this trip was to consolidate Sino-British military cooperation and expedite the Burma campaign, in addition to securing British commitment to China's sovereignty over Tibet.[62] Song's instincts were correct. When Churchill received Stanley Hornbeck, a special adviser to the US Department of State, in London in November 1943, he reiterated his stance over Hong Kong, along the lines of his Mansion House Speech in November 1942, that he was not going to be party to any "liquidation of the British Empire". But, according to Hornbeck, the Prime Minister was reported to have said that "perhaps some arrangement could be made with the Chinese whereby the question of sovereignty could be adjusted but the political control and administrative responsibility remain with Great Britain."[63]

At the Cairo Conference in November 1943 informal discussions on the issue of Hong Kong were held between Chiang Kai-shek and Roosevelt, and between Roosevelt and Churchill. At a meeting with Roosevelt, Chiang apparently appealed to the President to mediate between Britain and China on the issue.[64] According to Elliott Roosevelt's account, Roosevelt's precondition was that the Guomindang agree to organise a "unity government" with the Chinese Communists after the war.[65] Unfortunately, there is no indication of how Chiang responded to this idea. In hindsight, it is possible that Chiang accepted this proposal because at an ensuing meeting with Churchill Roosevelt was said to have put pressure on the British over the matter of Hong Kong. Churchill, however, refused any discussion of the colony's future.[66] In reinforcement of this, at

---

62 Ibid., 327–329.

63 Hornbeck's memorandum, November 15, on his conversation with Churchill, Hornbeck Papers, box 468, quoted in Chan, *China, Britain and Hong Kong, 1895–1945*, 312.

64 There is no official American record of the substance of this conversation. The only record is from the Chinese side. To compile *the FRUS*, the editor solicited the summary record in 1956 and then translated it into English under the title "Chinese Summary Record". See "Chinese Summary Record" of Roosevelt-Chiang Dinner Meeting (November 23, 1943), in US Department of State, ed., *Foreign Relations of the United States Diplomatic Papers, 1943, the Conference at Cairo and Tehran* (Washington, D. C.: Government Printing Office, 1961), 322–325. (hereafter *FRUS: 1943 Cairo and Tehran*) The record suggests that Roosevelt raised the matter of Hong Kong. But according to Liu Xiaoyuan, and another Chinese record, "Fifth Chinese Memorandum", it was Chiang who took the initiative. See Liu, *A Partnership for Disorder*, 130–139.

65 Elliott Roosevelt, *As He Saw It* (New York: Duell, Sloan, and Pearce, 1946), 164. Quoted from Liu, *A Partnership for Disorder*, 139.

66 Koo, *Gu Weijue huiyilu vol. 5*, 15–16; Chan, *China, Britain and Hong Kong, 1895–1945*, 313.

the Tehran Conference in December 1943 the Prime Minister proclaimed to Roosevelt and Joseph Stalin that "nothing would be taken away from England without a war", especially "Singapore and Hong Kong".[67] In March 1945, when Patrick Hurley, the US special envoy to China, made a further attempt to raise the issue with Churchill, the latter bluntly replied that "never would we yield an inch of territory that was under the British flag."[68] In response to Hurley's warning about the risk of "a complete nullification of the principles of the Atlantic Charter which was reaffirmed by Britain and the Soviet [Union] in the Iran Declaration", the Prime Minister's answer was that "Britain is not bound by the principles of the Atlantic Charter at all."[69]

In tandem with the US's failure to exert leverage over the British in relation to Hong Kong, China's enthusiasm for the colony was dampened by its military predicament and a deteriorating Sino-American relationship. In two memoranda of the Chinese Foreign Ministry in 1944, the free-port solution was no longer suggested; Chongqing sought to directly negotiate with British over the New Territories, together with other parts of Hong Kong, at an appropriate time.[70] This can be understood in a broader context. In 1944 China suffered calamitous reverses in the face of a massive Japanese attack: Operation Ichigo. As a result, China lost vast territories and more than half a million troops.[71] Chinese forces in the vicinity of Hong Kong retreated westwards into the mountains of Southwest China and thus lost a foothold to recapture Hong Kong. To compound matters, the United States became increasingly impatient with the perceived low morale and poor performance of the Nationalist troops, as well as alleged corruption and totalitarianism inside the Nationalist bureaucratic system. General George Marshall, the Chief of Staff of the US Army, even suggested to Roosevelt that Chiang Kai-

---

67 "Bohlen Minutes" of Tripartite Dinner Meeting (November 29, 1943), US Department of State, ed., *FRUS: 1943 Cairo and Tehran*, 552–554.

68 Quoted from Thorne, *Allies of a Kind*, 551.

69 Telegram from the Ambassador in China (Hurley), temporarily in Iran, to the Secretary of State, April 14, 1945, US Department of State, ed. *Foreign Relations of the United States Diplomatic Papers, 1945, the Far East, China*, vol. VII (Washington, D. C.: Government Printing Office, 1969), 329–332 (hereafter *FRUS: 1945 China*).

70 Waijiaobu memorandum, "Wancheng wo guo lingtu zhi wanzheng [Implement Our Country's Territorial Integration]", June 7, 1944, Victor Chi-ts'ai Hoo (Hu Shize 胡世澤) Papers, box 3, Hoover Institution, Stanford University, quoted from Liu, *A Partnership for Disorder*, 140; Waijiaobu 312.72, "Paper on methods to recover Hong Kong and the Kowloon Leased Territory", September 1944, quoted from Tsang, *Hong Kong: An Appointment with China*, 37.

71 Hsi-sheng Ch'i, "The Military Dimension, 1942–45," in *China's Bitter Victory: the War with Japan 1937–1945*, ed. James C. Hsiung and Steven I. Levine (New York and London: M. E. Sharpe, 1992), 165.

shek should be forced to hand over command of China's forces to General Joseph Stilwell, Chiang's US chief of staff (1942–1944), in the hope of reversing these battlefield setbacks. Chiang was infuriated by this humiliation and refused to agree unless Stilwell, with whom he had a stormy relationship, had been recalled. As a consequence, Sino-American relations were poisoned, even though Roosevelt soon overturned the decision and sent General Albert Wedemeyer to take over from Stilwell.[72]

Alongside this, there was a re-evaluation of China's likely contribution to Japan's defeat.[73] Marshall dismissed China's capacity for large-scale offensive against the Japanese and wished not to be drawn into a long campaign in China. He thus kept US military resources invested in East Asia at only the level necessary to keep up appearances.[74] As the US Navy advanced towards the Japanese home islands in the Pacific, the US military effort in China was reduced to "a sort of insurance policy ... if anything went wrong with the central Pacific offensive."[75] In the autumn of 1944, Japanese resistance against US amphibious landings in the Pacific was still fierce. Roosevelt urged the Soviet Union to shoulder some of the responsibility of defeating Japan. As an incentive, concessions, partly at the expense of China, were offered to Stalin at the Yalta Conference in February 1945. The President promised that Moscow might enjoy special privileges after the war in relation to two warm-water ports (Dalian for commercial use and Lüshun, or Port Arthur, as a Russian naval base), and two railways in Northeast China connecting the two ports to the Russian Far East, which were not divulged to the Chinese until May 1945.[76] These new developments did not escape Churchill's attention. He astutely perceived that "any claim by Russia for indemnity at the expense of China, would be favourable to our resolve about Hong Kong."[77]

In the meantime, the United States and Britain started to come to terms over some aspects of the colonial question. Washington became more divided on the matter as the war progressed. The Navy and Army Departments advocated full

---

**72** Mitter, *China's War with Japan 1937–1945*, 338–350; Hans van de Ven, "The Sino-Japanese War in History," in *The Battle for China: Essays on the Military History of the Sino-Japanese War of 1937–1945*, ed. Mark Peattie, Edward J. Drea and Hans van de Ven (Stanford, California: Stanford University Press, 2011), 450.
**73** Thorne, *Allies of a Kind*, 426–427.
**74** van de Ven, "The Sino-Japanese War in History," 451.
**75** Thorne, *Allies of a Kind*, 428.
**76** Liu, *A Partnership for Disorder*, 242–253.
**77** Churchill Archives (online), CHAR 20/153/4, WSC to Foreign Secretary and General Ismay, October 23, 1944, available at http://www.churchillarchive.com.ezproxy.is.ed.ac.uk/explore/page?id=CHAR%2020%2F153%2F4#image=9 (accessed April 7, 2019).

American control of the Japanese-mandated islands in the Pacific from the viewpoint of US security, but Roosevelt and the State Department worried that this might set a bad example to other powers.[78] To resolve this conundrum, the concept of a "strategic trust territory" was devised and discussed with the British. Through this mechanism, the Americans intended to assume sole administration over these Pacific territories, and the freedom to militarise them, but still pay homage to a central international trusteeship system that Roosevelt had always favoured.[79] Churchill preferred a US annexation of the Pacific islands but decided to use this as a bargaining chip to restrict the authority of the Trusteeship Council, one of the principal organs of the United Nations established to supervise the administration of non-self-governing territories. Indeed, at the San Francisco Conference in April-May 1945, the United States sided with Britain over the territorial scope of the International Trusteeship System. As a result, the trusteeship arrangement was limited to mandated territories and colonies of enemy countries, and "self-governance", as opposed to "independence" or "self-determination", for which the Soviet Russia and China strongly argued was set as the general goal of decolonisation.[80] Against this backdrop, and after Roosevelt's death in April 1945, President Harry S. Truman was not keen to press Britain on the matter of Hong Kong.

## 4 British Wartime Plans and Preparations for Restoring Hong Kong

British diplomatic circles, for example Seymour in Chongqing and Clarke in the Foreign Office, were not averse to discussions with the Chinese over the New Territories.[81] Gerald Gent, the Assistant Under-Secretary in the Colonial Office, pointed out that there might even be some room for Britain to compromise on the matter, using the early end of the lease as a bargain to exchange for joint control over the airport, reservoirs and other parts of the infrastructure in the New Territories essential for Hong Kong's well-being.[82] Nonetheless, mainstream opinion in London was to refuse any discussions over Hong Kong's future.[83]

---

**78** Louis, *Imperialism at Bay*, 448.
**79** Ibid., 522–523.
**80** Ibid., 530–541.
**81** Whitfield, *Hong Kong and the Anglo-American Alliance*, 91–93.
**82** TNA, CO 825/4215, memorandum of "Future Status of Hong Kong" by Edward James Gent, July 21, 1943. P1410337
**83** Whitfield, *Hong Kong and the Anglo-American Alliance*, 24–39.

Churchill's attitude was clear,[84] and when asked whether Hong Kong, or any other part of the Empire, was excluded from the Prime Minister's commitment at the Mansion House in 1942 not to preside over the liquidation of the British Empire, Clement Attlee, the leader of the Labour Party, also offered a reassurance that "no part of the British Empire or Commonwealth of Nations is excluded."[85]

But Churchill's and Attlee's resolve over Hong Kong could not deny the fact that the colony was far beyond British military reach at this time. Officials in London were nervous about the prospect of Chinese forces reoccupying Hong Kong after the defeat of Japan. In that scenario such troops "might prove difficult to extrude by ordinary diplomatic means."[86] The Colonial Office thus contrived to prevent this course of events and contemplated several plans to restore Hong Kong to British rule in advance of the Chinese forces.

## 4.1 The Hong Kong Planning Unit (HKPU)

Operating under the Colonial Office, the HKPU was an advisory and policymaking group for British post-war policy regarding Hong Kong. The cadres of the group would also constitute the core of the Hong Kong civil government once the colony was liberated. In February 1944 the unit had just nine members. It was only after David Mercer MacDougall, the former head of the Ministry of Information in Hong Kong, took over the unit in August of that year that it started to move onto a more solid footing.[87] The first task for MacDougall was to recruit more staff members. For this purpose, he organised a trip to East Asia in early 1945 and was shocked by the atmosphere that prevailed among the British communities east of Cairo. They were in deep doubt "whether Great Britain really and truly intended to take back and hold Hong Kong."[88] Many ex-Hong Kong civil servants, especially those of Chinese origin, had managed to travel to

---

**84** Churchill stated that "I have not become the King's First Minister in order to preside over the liquidation of the British Empire." See Churchill Archive (online), CHAR 9/189, "Prime Minister's Speech for Lord Mayor's Banquet [at the Mansion House]", November 10, 1942, http://www.churchillarchive.com.ezproxy.is.ed.ac.uk/explore/page?id=CHAR%209%2F189#image=70 (accessed April 7, 2019).

**85** TNA, FO 371/41657, parliamentary question by William Astor MP, November 8, 1944.

**86** TNA, HS 1/171, minute from Rait to DDMI (P/W), August 19, 1944. Quoted in Tsang, *A Modern History of Hong Kong*, 132.

**87** Whitfield, *Hong Kong and the Anglo-American Alliance*, 114.

**88** TNA, CO 129/592/8, letter from MacDougall to E.J. Gent, April 18, 1945.

Chongqing and were now working for the Chinese government.[89] MacDougall's plan was thus to set up a branch of the unit under the British embassy in Chongqing for recruiting purposes, but this encountered objections from Seymour. The latter was apprehensive about Chinese suspicions that "we were recruiting in China persons of Chinese race for the re-establishment of a British colony whose recovery is desired by China."[90] Consequently, the recruiting effort failed to get off the ground.[91] In this regard, MacDougall could do little to strengthen the HKPU except for approaching a few recruits in India and neutral Macao.[92]

MacDougall considered collaborating with the British Army Aiding Group (BAAG). The BAAG, which was made up of Chinese, British and other nationals who had lived in Hong Kong before the war, originally facilitated the escape and evacuation of allied POWs incarcerated in the colony but then extended its operations to collecting intelligence on the Japanese occupation of South China. It was an ideal group to aid MacDougall, but the head of the BAAG, Colonel Lindsay Ride (later Sir Lindsay Ride) declined his invitation. Ride, an Australian, had been on the teaching staff at the University of Hong Kong before enlisted into the army. During the war, he had been captured by the Japanese but had fortunately escaped and now aided other POWs. Ride cherished his close relationships with the Chinese. In fact, most of the BAAG's activities depended on its ethnic Chinese agents and amicable cooperation with Nationalist troops and the Communist guerrillas, the East River Column (*Dongjiang Zongdui* 東江縱隊), operating in the New Territories and their adjacent areas in Guangdong province such as Huizhou 惠州, Dongguan 東莞 and Bao'an. He was thus very concerned about possible Chinese protests.[93]

As an alternative, MacDougall planned to militarise the HKPU so as to play a leading role in the reoccupation of Hong Kong. However, the War Office was reluctant to accede to this. Even in August 1945, when Japanese unconditional surrender was close at hand, Brigadier French, the Deputy Director of Civil Affairs in the War Office, insisted that in Hong Kong there was no need "for all from the great bulk of the civil administration to be militarised." To him, it was imprac-

---

**89** TNA, FO 371/46251, draft telegram from MacDougall to Mr Sansom (the China Office at Calcutta), May 3, 1945.
**90** TNA, FO 371/46251, telegram from Chongqing to Foreign Office, May 18, 1945.
**91** In July 1945 the plan was still under discussion in the Colonial Office. See TNA, CO 129/591/14, "copy of minute (original on 54144/25/34)", July 20, 1945.
**92** TNA, FO 371/46251, letter from E.J. Gent (CO) to J.C. Sterndale Bennett (FO), June 22, 1945.
**93** TNA, CO 129/591/11, "extract from Mr MacDougall's report on visit to [Chongqing]", May 3, 1945.

tical for a group of some 40 officers to occupy Hong Kong. Even it could have achieved some success, this would not have done much to enhance the British Empire's reputation.[94] In the end, MacDougall did not fulfil his desire to play a prominent part in the reoccupation of Hong Kong, but substantial preparations were made for the restoration of the British Hong Kong administration.

## 4.2 The British Response in the Case of Hong Kong Being Occupied by American or Chinese Troops

Given the main focus of the HKPU was on the reestablishment of the Hong Kong civil administration, the Colonial Office also had to make contingencies in case Hong Kong was occupied by the Chinese or Americans before any British forces arrived in the colony. As wartime military cooperation between Britain and the United States deepened, some understanding with Washington had been achieved on the issue of post-war military occupation in September 1943. "The Charter for the Combined Civil Affairs Committee at Washington" provided that

> When an enemy occupied territory of the United States, the United Kingdom or one of the Dominions is to be recovered as a result of an operation, combined or otherwise, the military directive to be given to the force commanders concerned will include policies to be followed in the handling of civil affairs as formulated by the government which exercised authority over the territory before enemy occupation. If paramount military requirements, as determined by the Force Commander, necessitate the departure from those policies, he will take action and report through the Chiefs of Staff to the Combined Chiefs of Staff.[95]

The agreement only concerned "policies" and it was unclear if this carried the implication that a British civil administration could be restored. However, the Americans tacitly approved this in the case of Borneo. A British Borneo Unit, assembled in Australia since January 1945, was thus permitted by the South-West Pacific Command (SWPC) to re-establish British rule. The Colonial Office desired a similar arrangement for Hong Kong.[96] But the situation in Borneo was different to that in Hong Kong. The arrangement in Borneo was partly a result of the Brit-

---

**94** TNA, CO 129/591/16, minute from E.J. Gent (CO) to George Gater (CO), August 10, 1945.
**95** TNA, CO 129/591/13, "The Charter of the Combined Civil Affairs Committee at Washington" (under the co-Chairmanship of British and American representatives) was approved by the Combined Chiefs of Staff in September 1943, attached to the telegram from G.H. Gater (CO) to Alexander Cadogan (FO), July 28, 1945.
**96** TNA, CO 129/591/15, memorandum of "Future Operations in China", attached to the letter to Major F.J.T. Durie, August 10, 1945.

ish being on good terms with General Douglas MacArthur, Supreme Commander in the SWPC. His strained relations with his own country's intelligence agency, the Office of Strategic Services (OSS), inadvertently led him to rely more on its British counterpart, Special Operations Executive Australia.[97] The British, however, did not have such luck with the China Command.

### 4.2.1 The Emergency Administration Plan

An Emergency Administration Plan was contemplated by the Colonial Office and the War Office in early 1945 in the event that Hong Kong was seized by Chinese troops, whether regular or irregular. In view of the possible withdrawal of Japanese troops from Hong Kong and their move to a more defensible location from late 1944, there was a real possibility that the Chinese regular army, or some Chinese warlord or commander of an irregular band looking for plunder or glory, would exploit the situation and it would be a diplomatic disaster for the British to have to try and grab Hong Kong back from an allied force. By the same token, it would be an embarrassment if the British had to plead with the Chinese to return the colony. With these scenarios in mind, Oliver Stanley, the Colonial Secretary, attempted to stake out some countermeasures in collaboration with the War Office.

In a letter to James Grigg, the Secretary of State for War, Stanley recommended the establishment of a Hong Kong emergency unit which could quickly enter Hong Kong and forestall any possible seizure by the Chinese if a military vacuum occurred in the colony. The unit would be deployed "as near the spot as possible" and "assume the responsibility for the civil government until the arrival of the regular civil affairs administration." In Stanley's mind, the BAAG was the ideal force to form the nucleus of such a unit because Colonel Ride had a deep understanding of Hong Kong conditions.[98] Grigg did not want the War Office to become closely involved in this project but agreed to provide some military help. He emphasised that the "emergency administration" cannot be "a Military Government or a Civil Affairs Administration based on military sanctions." It would be necessary for the Colonial Office to organise the administration.[99] A suitable instruction was thus drafted for General Hayes, the British Military Attaché to China, who commanded all British forces in China, the BAAG included.[100]

---

97 TNA, CO 129/591/11, letter to G. Gater, March 19, 1945.
98 TNA, CO 129/591/11, letter from Oliver Stanley to James Grigg, January 23, 1945.
99 TNA, CO 129/591/11, letter from James Grigg to Oliver Stanley, February 12, 1945.
100 TNA, CO 129/591/11, draft telegram from the War Office to SACSEA, annexed to the telegram from G.E.J. Gent (CO) to B.F. Montgomery, April 19, 1945.

However, to the surprise of British officials, Ride declined to involve his unit in the operations to reoccupy Hong Kong. In April 1945 he telegraphed General Hayes arguing that

> If the BAAG is used in the operations on Hong Kong for any purpose outside its official role (escape and evasion of landing parties and airmen and rescue of P/W and security) we shall be damned in the eyes of both Chinese and Americans for a long time afterwards. The comment will be 'There you are; we knew all along the perfidious British were up to their old games.'[101]

As a concession, Ride would only permit those with loose connections to the Chinese to join a Civil Affairs Unit. This excluded himself and most other cadres of the BAAG and thus gave little assistance to this plan.[102] Another objection came, predictably, from General Wedemeyer, the US Chief of Staff to Chiang Kai-shek.[103] He demanded that any para-military operations in the China Command areas should be approved by himself and this included the activities of the BAAG. He further underlined that it was his policy to "approve no para-military activities which he believed would be contrary to the wishes of the Generalissimo."[104] In this situation, George Gater, the Permanent Under-Secretary of the Colonial Office, decided to defer any decision on the emergency unit.[105]

The Colonial Office considered using the Special Operations Executive (SOE) as an alternative to the BAAG.[106] However, the SOE had little presence in China due to Chinese suspicions and competition from the American OSS. The first joint project, the China Commando Group, involving SOE and the Chinese government, only lasted eight months.[107] Thereafter, the SOE had to share the clandestine networks of its Chinese counterparts, at first the Institute of International Relations (IIR) and then the International Intelligence Service (IIS).[108] It was not until the end of 1943 that SOE was authorised to have its first office in China,

---

**101** TNA, CO 129/591/11, "extract from a report by Colonel Ride to B.M.A., [Chongqing]", April 25, 1945.
**102** Ibid.
**103** TNA, CO 129/591/11, letter from Frederick Bovenschen to George Gater, March 22, 1945.
**104** TNA, CO 129/591/11, minute from E.J. Gent to G. Gater, March 6, 1945.
**105** TNA, CO 129/591/11, letter from George Gater to Frederick Bovenschen, March 26, 1945.
**106** TNA, CO 129/591/11, minute from E.J. Gent to G. Gater, March 6, 1945.
**107** Maochun Yu, "'In God We Trusted, In China We Busted': the China Commando Group of the Special Operations Executive (SOE)," *Intelligence and National Security* 16, no. 4 (2001): 37.
**108** IIR operated in the vicinity of Hong Kong but it was a body directly responsible to Chiang Kai-shek who had no wish for British-sponsored sabotage. IIS's main operation zone was in North China. See Charles Cruickshank, *S.O.E. in the Far East* (Oxford and New York: Oxford University Press, 1983), 151–152.

camouflaged as the office of the Assistant Military Attaché in Kunming. It then established a branch in Guilin 桂林, the China Coast Section (CCS), using the BAAG as its cover, and one of the CCS's duties was to carry out future special operations on the South China Coast within the vicinity of Hong Kong. But little had been achieved by the time Guilin fell into the hands of the Japanese in late 1944.[109] After that, the CCS was reformed as "Group C" of the Indian Mission with headquarters in Kunming, reporting directly to India. The Indian Mission had envisaged some operations in the areas surrounding Hong Kong and an important element of these plans was the creation of a Sino-British Resistance Group – 30,000 guerrillas to be raised in Guangxi and Guangdong Provinces, as a major force to recapture Hong Kong. Yet Wedemeyer regarded the British operatives as interlopers from the British South East Asia Command (SEAC), and he subsequently forbade any British schemes in his command unless approved by himself.[110] As a result, there was no room in the China theatre for the SOE to play an active role.

With little prospect of SOE's involvement, there was a return to the BAAG option in July 1945. Gent was opposed to dropping Ride as he believed the latter's "experience and capabilities were too valuable to dispense with."[111] This was a view shared by John C. Sterndale Bennett, the then head of the Far Eastern Department of the Foreign Office (1940–1942 and 1944–1946). In contrast to Ashley Clarke, Sterndale Bennett held a much stronger stance on the recovery of Hong Kong.[112] On July 5, an inter-departmental meeting between the Foreign Office and the Colonial Office resolved that it was the duty of the BAAG to serve British interests in East Asia, namely the recovery of Hong Kong, regardless of Ride's personal feelings. The Foreign Office was assigned to change his mind.[113] Yet in view of Wedemeyer's likely obstruction, Seymour deemed the plan impractical and so did General Hayes, the Military Attaché, and General Carton de Wiart, Churchill's private representative in China. In a telegram to the War Office, General Hayes maintained that

---

**109** The ineffectiveness also derived from Colonel Ride, the commander of the BAAG, who vetoed any SOE activities which seemed to interrupt his own plans or jeopardise his own organisation. See ibid., 153–158. The SOE attempted a merger with BAAG, but this was rejected by the Military Attaché on the grounds that BAAG was the only British force in China with any creditable reputation, and that it would likely lose that reputation if it was amalgamated with SOE. See TNA, CO 129/591/11, minute to G. Gater, March 19, 1945.

**110** Cruickshank, *S.O.E. in the Far East*, 160–162.

**111** TNA, CO 129/591/10, minute by Ms Ruston, June 22, 1945.

**112** Tsang, *Hong Kong: An Appointment with China*, 41–42.

**113** TNA, CO 129/591/11 and FO 371/46251, "Note of a Meeting Held to Consider the Questions Raised in Colonel Ride's report of the 25th April", July 5, 1945.

[The Americans] are in any case extremely suspicious of our activities ... Hitherto these difficulties have been mitigated by a policy of complete frankness with Wedemeyer on the part of General Carton de Wiart and myself. We are both completely committed to keeping him informed of all (repeat all) we do in the military field. ... We both have gained the impression that Wedemeyer is personally opposed to any action by the British in China directed to the reoccupation of Hong Kong.[114]

In Hayes's mind, the only option was to secure permission directly at the highest levels in China and America to proceed with emergency planning so as to circumvent General Wedemeyer. In the end, the "Emergency Administration Plan" was cancelled by the War Office after nearly seven months of contemplation.[115] At the end of July 1945 it was agreed across Whitehall that if a British operation to recover Hong Kong had any chance of success, it had to be arranged from outside China.[116]

### 4.2.2 The Military Liaison Mission Plan

In contrast to the Emergency Administration Plan, which dealt with the anticipated seizure of Hong Kong by Chinese forces after the Japanese evacuation, the Military Liaison Mission Plan was a response to a US amphibious campaign plan that would be carried out after the seizing of Okinawa, the main island of the Ryukyu Islands. The aim of the US plan was to open up Chinese coastal cities by an attack from the sea, in company with simultaneous land offensives by Chinese forces from the west. In April 1945 Brigadier General Olmsted, Chief of the G-5 section of the United States Army Headquarters in China, asked for British collaboration.[117] He hoped that the British would provide a liaison officer for consultation over the utilisation of British properties and installations in these seaboard territories.[118] This offered a precious opportunity for the British to exert their influence. Seymour thus suggested a full liaison mission within the

---

114 TNA, CO 129/591/13 and FO371/46251, telegram from Britmis Chongqing to War Office, July 20, 1945.
115 TNA, CO 129/591/11, telegram from War Office to BRITCHIN, August 3, 1945.
116 TNA, CO 129/591/11, "Note of a Meeting held in the Colonial Office to consider the general question of the circumstances in which Hong Kong is likely to be re-taken and in particular the points raised in telegram No. 0/0114 of the 20th July from the G.O.C. British Troops in China", July 23, 1945.
117 The G-5 section is an American institution attached to the Command of the China Theatre, which was responsible for civil (Economic and Political) affairs, Lend-Lease operations, training of allied military forces and clandestine operations.
118 TNA, CO 129/591/14, telegram from H. Seymour (Chongqing) to Foreign Officer, April 11, 1945.

Chinese military headquarters as its forces advanced. This mission's task would be to gather information about "British properties and facilities in areas to be liberated and personnel who might be able to look after British subjects and property as and when liberated." In parallel with this, similar liaison arrangements should be made with "sea-borne invading forces whenever the landing may be made." Here was a chance to shape events in Hong Kong.[119]

Seymour's suggestions were welcomed by the Colonial Office as the mission could be used as "an umbrella" to "shelter the Hong Kong officers."[120] However, this promising endeavour soon proved to be groundless since the Americans and the Chinese changed their strategy. Wedemeyer was now contemplating a military campaign towards Guangzhou and Hong Kong by Chinese forces trained by American officers.

## 5 The Race for Hong Kong and Sino-British Disputes over the Right to Accept the Japanese Surrender in Hong Kong

### 5.1 Wedemeyer's Strategy for Guangzhou and Hong Kong

The campaign towards Guangzhou and Hong Kong was aimed at opening an ocean supply line to China and had been under consideration since early 1945.[121] In March, Chiang Kai-shek had earmarked 13 corps, a total 39 divisions, fully trained and equipped by the Americans, for these operations. By July some progress had been achieved, with Nanning 南寧 liberated, and a new push about to start towards the Leizhou 雷州 peninsular (Guangzhouwan). Guangzhou, it was anticipated, would be recovered in November, followed by Hong Kong.[122]

Wedemeyer's plan for Hong Kong was kept secret from the British. London only found out about it during the Potsdam Conference (July 17-August 2, 1945, codename "Terminal") through information provided by the US Chiefs-of-

---

**119** TNA, CO 129/591/14, telegram from H. Seymour (Chongqing) to Foreign Office, April 17, 1945.
**120** TNA, CO 129/591/14, minute by Miss A. Ruston, May 28, 1945, and minute by Miss A. Ruston, July 16, 1945.
**121** Chongxi Bai, *Bai Chongxi huiyilu [The Reminiscences of General Pai Chung-hsi]* (Beijing: Jiefangjun Chubanshe, 1987), 225 – 226.
**122** S. Woodburn Kirby, *The War Against Japan Volume V: the Surrender of Japan* (London: HMSO, 1969), 139, 144. Quoted in Tsang, *Hong Kong: An Appointment with China*, 38.

Staff.[123] On August 3, the details reached the British through the US Pacific Command rather than through the China Command.[124] This was because Wedemeyer's plan required the collaboration of the US fleet in the Pacific. It is possible that some American naval officials might have differing attitudes to their colleagues in China regarding any British reoccupation of Hong Kong.

Wedemeyer's operation galvanised Whitehall into action. On July 24, a memorandum, entitled "Arrangements for the administration of Hong Kong in the event of its liberation by regular Chinese forces", was drafted. It suggested the British government to obtain an agreement from Chiang Kai-shek that "from the outset of any initial period of military administration our Civil Affairs policies should be accepted, and that the Civil Affairs administration should be conducted by British personnel."[125] But London was more inclined to bypass Chiang and Wedemeyer. The preferred solution for the Foreign and Colonial Offices was thus "squaring the Americans in advance of an approach to Chiang Kai-shek." To save time, it was proposed that the Prime Minister should immediately discuss this with President Truman at the Potsdam Conference.[126] However, slowed down by the ongoing transfer of power from the Conservative Party to the Labour Party after the 1945 British general election, the Prime Minister and the Foreign Secretary preferred to use the "usual [diplomatic] channels" to proceed with the issue.[127]

Even so, it was very possible that, according to Wedemeyer's plan, Hong Kong would be liberated by Chinese troops. In early August, Chinese forces had captured Wuzhou 梧州 and were approaching the border of Guangdong province.[128] Of course, if Hong Kong was occupied by the Chinese, this did not necessarily mean that Chiang would hold on to the colony and refuse to return it to Britain since he had agreed to deal with the issue through diplomatic negotiations. But it would indeed give him strong leverage in any ensuing discussions over Hong Kong. However, history was unpredictable. Only a few days later, the Japanese chose to capitulate after the Americans dropped two nuclear bombs on

---

123 TNA, FO 371/46251, draft minute to Prime Minister, titled as "Arrangements for the Administration of Hong Kong in the event of its liberation by regular Chinese forces", July 24, 1945, and letter from J.C. Sterndale Bennett to L.H. Foulds, TERMINAL, July 28, 1945.
124 TNA, CO 129/591/15, telegram from C. in C. B.P.F. to Admiralty, August 3, 1945.
125 TNA, FO 371/46251, draft minute to Prime Minister, titled as "Arrangements for the Administration of Hong Kong in the event of its liberation by regular Chinese forces", July 24, 1945.
126 TNA, FO 371/46251, minute by J.C. Sterndale Bennett, July 25, 1945; TNA, FO 371/46251, letter from J.C. Sterndale Bennett to L.H. Foulds, TERMINAL, July 28, 1945.
127 TNA, CO 129/591/13, letter from J.C. Sterndale Bennett to G.E.J. Gent, August 10, 1945.
128 Tsang, *Hong Kong: An Appointment with China*, 38.

Hiroshima and Nagasaki on August 6 and 9. This ended Wedemeyer's plan and provided a chance for the British to reoccupy Hong Kong on their terms.

## 5.2 Arguments over the Right to Accept the Japanese Surrender

In July 1945, the Japanese troops had still appeared tenacious and unyielding. No one could anticipate the scenario that Japan would surrender within a few weeks. In early August, the British Chief of the Imperial General Staff still believed that Wedemeyer's forthcoming operation in Zhanjiang 湛江 (Fort Bayard) "was unlikely to lead to the rapid capture of Hong Kong, since the Japanese opposition in that area was very strong."[129] However, the circumstances dramatically changed in the wake of the atomic bombs and the Soviet Union's campaign against Japan.

On August 10, the Chiefs of Staff had in preparation a plan for the disposition of British military forces in the event of a sudden surrender by Japan. A British occupation of Hong Kong was included. "The most convenient course," it was suggested, "would probably be to send a naval force to Hong Kong with a detachment of Marines" and "this force would probably be drawn from the British Pacific Fleet." Clement Attlee, the new Prime Minister, would send "a personal telegram to President Truman on this point."[130] After the liberation, "a period of military administration was essential in Hong Kong." Concurrently, "MacDougall and his civilian unit in England should be immediately mobilised and despatched to the Far East" in order to accompany the naval force to oversee civil affairs, and subsequently form a civilian government as a replacement for the military one.[131] If Franklin Gimson, the former Colonial Secretary of Hong Kong, and other senior officials were released from internment camps before the British force arrived, they were encouraged to restore the British administration in Hong Kong immediately. Indeed, Gimson was to be given authority to administer Hong Kong "in [the] absence of [the] Governor under existing Letters Patent."[132]

Meanwhile, the British were in the dark about Chinese plans to receive the Japanese surrender.[133] Officially, Chiang Kai-shek, as the Supreme Commander

**129** TNA, CO 129/591/15, minutes of Chiefs of Staff Committee, Joint Planning Staff, titled "Future Operations from China", August 8, 1945.

**130** TNA, CO 129/591/16, "extract from minutes of a Cabinet Meeting", held on August 10, 1945.

**131** TNA, CO 129/591/16, minute from E.J. Gent to G. Gater, August 10, 1945.

**132** TNA, CO 129/591/16, telegram from Foreign Office to Chongqing, August 11, 1945.

**133** TNA, CO 129/591/16, telegram from BAAG to BT China, August 15, 1945.

in the China Command areas, had authority to accept the Japanese surrender in Hong Kong, verified in the later US General Order No. 1 for the surrender of Japan. Chiang thus designated the Chinese Thirteenth Corps, under Shi Jue 石覺, to take over Hong Kong while the New First Corps, led by Sun Liren 孫立人, headed for Guangzhou. The two military forces were in the vicinity of Wuzhou, numbering more than 60,000 and with US equipment and materials.[134] However, before General Order No. 1 was formally approved by President Truman on August 17, 1945, other Anglo-American arrangements had been made. Indeed, the United States Chiefs of Staff had given the British Pacific Fleet, which was at the time under the control of the United States Pacific Fleet, permission to accept the Japanese surrender in Hong Kong.[135] Although the US Chiefs of Staff emphasised that "the release of British Forces is viewed by the United States Chiefs of Staff as being unrelated to any proposals of the British Chiefs of Staff concerning Hong Kong", their stance could not have been clearer.[136] As part of these arrangements, Chester Nimitz, Commanding Officer of the United States Pacific Fleet, was ordered to provide the necessary support to the British, including mine sweeping and oil supplies.[137] The British decided to keep Chiang Kai-shek and Wedemeyer in the dark: "There should be no question of consultation with the Chinese about the reoccupation of a British possession, but that the Chinese Government should merely be informed of what we are doing a few hours before our forces are due at Hong Kong."[138]

However, concealment quickly proved impracticable because of the potential clashes with the putative surrender plans of Wedemeyer.[139] The British Embassy in China thus formally informed Chiang on August 16, 1945.[140] Wedemeyer,

---

134 Tsang, *Hong Kong: An Appointment with China*, 45–46.

135 TNA, FO 371/46252, letter from Sterndale Bennett (FO) to Colonel G.R. Price (Cabinet Offices), August 14, 1945.

136 TNA, FO 371/46252, telegram (141625 A) from Admiralty to C. in C.B.P.F., August 14, 1945.

137 TNA, FO 371/46252, telegram from C. in C.B.P.F. to C.T.F. III., August 15, 1945.

138 TNA, FO 371/46252, letter from Colonel G.R. Price (Cabinet Offices) to Sterndale Bennett (FO), August 14, 1945.

139 Wedemeyer had contemplated his own surrender plan for Hong Kong and had ratified it on August 15. See TNA, FO 371/46252, telegram from S.A.C.S.E.A to A.M.S.S.O, August 15, 1945.

140 TNA, CO 129/591/16, telegram from Sir H. Seymour (Chongqing) to Foreign Office, August 16, 1945. The fact provided by Seymour in the telegram that Chiang was informed of the Hong Kong scheme on "12 August" is incorrect, because at that time London had not solicited the US permission yet. Illustrated by other British archival documents, the date when Ambassador was ordered by the Foreign Office to notify Chiang was August 15, 1945 (see TNA, CO 129/591/16, telegram of No. 888 from Foreign Office to Chongqing, August 14, 1945; TNA, FO 371/46252, telegram of No. 890 from Foreign Office to Chongqing, August 14, 1945; and TNA, FO 371/46252, telegram of No. 891 from Foreign Office to Chongqing, August 15, 1945). According to Chinese sources,

meanwhile, received a message from Washington that his task in Hong Kong was merely to assist the evacuation of internees and prisoners.[141] The news enraged Chiang Kai-shek since he could not bear that his authority as the supreme commander of China Command had been usurped over Hong Kong.[142] Yet, as regards the sovereignty of Hong Kong, Chiang was not looking to pick quarrels and made this clear in his reply to the British:

> China has no territorial ambitions either in Hong Kong or French Indo-China and would not take advantage of the fact that these territories are in China theatre to establish possession. The Chinese Government regarded Hong Kong as a matter which would require eventual settlement through [a] diplomatic channel.[143]

Chiang thus ordered that there should be no "race with Britain for entering Hong Kong but the Chinese troops would occupy the New Territories if the time permitted."[144]

In parallel with their disputes with the British over Hong Kong, the Nationalists faced intense internal competition from the Communist Party. The sudden Japanese capitulation disrupted Chiang and Wedemeyer's military strategy in China. In their estimation, one or two years would be needed to begin a series of full-strength counterattacks so as to recapture Chinese territories from Japanese forces city by city. The sudden surrender left a huge political vacuum in China proper for the Chinese Nationalist Party and the Chinese Communist Party to compete to fill. At this time, the majority of the Nationalist troops were trapped in the mountains of Southwest China after their military defeat by the Japanese during the Ichigo operation, while the Chinese Communist forces, though much weaker, were better placed to fill the void. With the help of their guerrillas in the zones of occupation the Communist forces rushed to take over as much territory as possible from the Japanese. As far as Chiang was concerned, this competition with the Communist Party was vital to the sur-

---

Chiang had been aware of the British intention in regard to Hong Kong from American channels but received the official British notification at noon on August 16 (see Zhenghua Wang, ed., *Jiang Zhongzheng zongtong dang'an: shilüe gaoben [Draft Papers of Chiang Kai-shek]*, 83 vols., vol. 62 (Taipei: Academia Historica, 2011), 203–206 (hereafter cited as *SLGB vol. 62.*).

**141** TNA, CO 129/591/16, Telegram from Mr. Wallinger (Chongqing) to Foreign Office, August 16, 1945.

**142** TNA, CO 129/591/16, telegraph (No. 856) from Sir H. Seymour (Chongqing) to Foreign Office, August 16, 1945.

**143** TNA, CO 129/591/16, telegraph (No. 857) from Sir H. Seymour (Chongqing) to Foreign Office, August 16, 1945.

**144** August 15, 1945, Wang, ed., *SLGB vol. 62*, 203.

vival of his regime. In contrast, the rivalry with Britain over the matter of Hong Kong was of less urgency. In fact, he might need British cooperation to facilitate the transportation of his troops northward by using British ships and the modern port facilities in Hong Kong. This perhaps explains why he held back the Chinese forces when they had already advanced to the borders of Hong Kong before the British fleet's arrival.[145]

Chiang's priorities were correctly perceived by Seymour, who believed the core of the dispute over the surrender was not the future of Hong Kong but China's "face". He thus suggested a compromise which would allow Chiang's representative to "accept surrender jointly with Flag Officer of British Naval Force."[146] However, Whitehall's attitude was high-handed. It was widely believed in London that the British had the right to "accept Japanese surrender in its own territory" irrespective of operational theatres and London's agreement with Washington antedated any terms of General Order No. 1.[147]

Attlee decided to seek American endorsement of this.[148] Truman's response was favourable to the British. He indicated that "there is no objection [from the United States] to the surrender of Hong Kong being accepted by a British officer" on the condition that "full military coordination is effected beforehand by the British with the Generalissimo."[149] Nonetheless, Truman sought to strike a balance. He informed Song Ziwen, now the Chinese premier, that his decision over the surrender arrangements did not in any way represent the United States' views regarding the future status of Hong Kong.[150] Moreover, in his response to the protests of Chiang he stressed that "no question arises with regard to British sovereignty in the area" and "[whether] it is practicable to do so, surrender by Japanese forces should be to the authorities of that nation exercising sovereignty in the area."[151]

Without America's support, Chiang Kai-shek was doomed to yield. On August 30, a combined British force, including the British Pacific Fleet, a R.A.F. airfield construction force, and a commando brigade from the SEAC, arrived in

---

**145** On August 28, Chiang instructed his Chinese troops *en route* to the New Territories to stop. See ibid., 367.

**146** TNA, CO 129/591/16, telegraph (No. 857) from Sir H. Seymour (Chongqing) to Foreign Office, August 16, 1945.

**147** TNA, CO 129/591/16, telegraph from Foreign Office to Chongqing, August 17, 1945.

**148** TNA, FO 371/46252, telegraph from Prime Minister to President Truman, August 18, 1945.

**149** TNA, CO 129/591/16, telegraph from President Truman to Prime Minister, August 18, 1945.

**150** The message was relayed by an American telegram to Britain, see ibid.

**151** Telegram from Secretary of State to the Ambassador in China (Hurley), August 21, 1945, US Department of State, ed., *FRUS: 1945 China*, 509.

Hong Kong. Before its arrival, Franklin Gimson, now freed, had already assumed temporary authority after receiving authorisation relayed by the BAAG.[152] Mac-Dougall and eight key members of the HKPU landed in Hong Kong on September 7. Soon they set about rebuilding the civil administration in the colony.

In the end, some efforts were made to save Chiang's face. The surrender ceremony in Hong Kong was held on September 16, seven days after the equivalent ceremony at Nanjing. Admiral Cecil Harcourt, as both Britain's and the China Theatre's representative, accepted Japanese surrender. Though curious about the American request for "full military coordination", the British decided to give the Generalissimo assurances that "so far as the conditions of the port allow, full facilities through the Hong Kong area" would be given "for the assistance and support of Chinese and American forces who may still be either engaged against the enemy or involved in securing the surrender of Japanese forces in the hinterland."[153] Accordingly, a coordination committee was established to deal with the utilisation of facilities for the embarkation of Chinese troops for redeployment; the transfer of all Japanese ships, aircraft and other weapons and equipment captured in Hong Kong to China; and assistance in investigating alleged war criminals and collaborators who might have taken refuge in Hong Kong.[154]

## Conclusion

The Sino-British competition for Hong Kong during the Pacific War was sparked by the Nationalist attempt to connect the matter of the New Territories to the abolition of British extraterritoriality in the 1943 treaty negotiations. The Nationalists believed that it would be a good start to recover the whole of Hong Kong. However, this proposal was quickly rebuffed by the British and received little American support, partly because the Nationalists had selected the wrong ground on which to fight. The New Territories, although leased to Britain, had no similarities with settlements and concessions where British rights were based on the treaty-port system and a wider interpretation of extraterritoriality.

---

152 TNA, CO 129/591/16, telegram from Mr. Wallinger (Chongqing) to Foreign Office, August 14, 1945; Edwin Ride, *B.A.A.G., Hong Kong Resistance, 1942–1945* (Hong Kong: Oxford University Press, 1981), 292.
153 TNA, FO 371/46252, telegram from Foreign Office to Chongqing, August 21, 1945.
154 TNA, CO 129/592/19, telegram from BRITCHIN to A.M.S.S.O. and Troopers, August 30, 1945, and telegram from M.A. China to A.M.S.S.O., September 1, 1945.

As a compromise, Chongqing and London eventually agreed to shelve the question until the war ended.

This postponement, however, created uncertainties as to Hong Kong's future, coupled with doubts as to whether China or Britain would reoccupy the city from the Japanese forces. In 1943–1944 the Nationalist strategy was to force the British into returning the colony to China by enlisting the Americans' endorsement. However, this approach failed due to the vehement opposition of the British, particularly on the part of Churchill, and American half-heartedness. London, in the end, managed to ease the Hong Kong topic off the Allies' international agenda. Meanwhile, the Colonial Office contemplated a series of contingency plans in the event that the colony was first occupied by Chinese forces, but none of these was put into practice mainly because of the lack of British military resources in East Asia. In this period, China's ambition to recapture Hong Kong was to some degree exaggerated by the officials in London. Unbeknownst to them, the Chinese enthusiasm for Hong Kong had already cooled when Sino-American relations encountered problems.

The British reoccupation of Hong Kong can largely be attributed to the sudden Japanese capitulation. The Americans played a prominent role in facilitating this return. But the main American actors were not necessarily working to the same script. While US officials in China attempted to hamper the British in their activities regarding Hong Kong, their seniors in Washington tended to facilitate British claims over the colony. The high-level wartime military cooperation between Britain and the United States, the Anglo-American conciliation over colonial matters, as well as the cooling of the Sino-American relationship towards the end of the war, all contributed to this result.

# Chapter IV
# The Unsettled Future of Hong Kong, 1946–1949

This chapter explores Sino-British interactions over Hong Kong's future during the period of 1946–1949. As previously indicated, the sovereignty of the New Territories had not been resolved by the two countries in wartime. Yet Britain had committed itself to discuss the lease "when victory is won".[1] At the end of the conflict, the question of Hong Kong naturally re-emerged. But how high this issue was on the Nationalist government's agenda merits consideration.[2] As this chapter will demonstrate, the Chinese government's view differed from that of the Chinese public on the immediate reversion of Hong Kong to China. Meanwhile, the British government was formulating policy over Hong Kong's future and had to decide how far it was prepared to compromise over sovereignty. Local Hong Kong opinion was pertinent here. In the end, the question of Hong Kong was left unsettled before the Nationalist authorities retreated to Taiwan in 1949. This chapter traces the complicated twists and turns of the two protagonists during this period.

## 1 The Nationalist Attitude towards the Recovery of Hong Kong, 1945–1946

The Guomindang's enthusiasm for Hong Kong had faltered in the latter stages of the war. The Nationalist regime was preoccupied with more urgent issues. Domestically, the Guomindang was in competition with the Chinese Communist Party (CCP) to resume Chinese administration over the occupied areas, particularly in North China and Northeast China, where the CCP's influence had expanded quickly after the Japanese advanced southward during Operation Ichigo.

---

1 TNA, CO 129/588/24, telegram from Sir H. Seymour to Foreign Office, December 25, 1942; Best, ed., *BDFA, Part III, Series E, vol. 6*, no. 49, 95.
2 Many historical works have not focused on this matter, such as Chan, *China, Britain and Hong Kong, 1895–1945*; ibid., "Hong Kong in Sino-British Diplomacy, 1926–45"; Catherine R. Schenk, *Hong Kong as an International Financial Centre: Emergence and Development 1945–65* (London and New York: Routledge, 2001); Whitfield, *Hong Kong and the Anglo-American Alliance*; Steve Tsang, *Governing Hong Kong: Administrative Officers from the Nineteenth Century to the Handover to China, 1862–1997* (London and New York: I. B. Tauris & Co Ltd, 2007); ibid., *Hong Kong: An Appointment with China*; ibid., *A Modern History of Hong Kong*; Wesley-Smith, *Unequal Treaty 1898–1997*; Tang, "World War to Cold War: Hong Kong's Future and Anglo-Chinese Interactions, 1941–55."

https://doi.org/10.1515/9783110706659-011

Externally, Nationalist energy was concentrated on enlisting US support against the CCP and facilitating an early withdrawal of Soviet troops from Northeast China and the return of these territories to its control. As an outcome of the Yalta Conference, the Soviet Union had entered the war against Japan in August 1945 and had subsequently occupied Northeast China. It was widely supposed by the Nationalists that the Soviet occupation had assisted the CCP's development and infiltration into these territories.

As a result of such pressures, the question of Hong Kong was set aside for a while. In a statement by Chiang Kai-shek on August 24, 1945, it was noted that "the present status of Hong Kong is regulated by a treaty signed by China and Great Britain"; "changes in future will be introduced only through friendly negotiations between the two countries."[3] Instead of resuming the Sino-British argument over the colony, Chiang at this time gave more importance to London's role in helping to redeploy Chinese troops trapped in the mountains of Southwest China to North China through its modern port facilities. Indeed, this was a promise that the British had made to solicit China's agreement over the Japanese surrender arrangements in Hong Kong.[4] Besides this, the British had also agreed to a joint Sino-British effort to prevent war criminals taking refuge in Hong Kong and to the turning over to the Nationalists the Japanese ships, weapons and equipment captured in Hong Kong. During this period the Chinese 1st, 8th, 13th, 54th and 93rd Armies were transported by American ships from Hong Kong to North China,[5] and a large district in Hong Kong, known as Jiulongtang 九龍塘 (Kowloon Tong), was set aside to accommodate these troops.[6]

Yet in contrast to the Nationalist government's pragmatic attitude, the Chinese public clamoured for the immediate return of Hong Kong. The Nationalists thus endeavoured to find a way to placate Chinese public opinion but at the same time avoid any formal discussions with the British over the issue. As a result, the Nationalist stance towards Hong Kong appeared strong on rhetoric but weak on action. This characteristic is illustrated by the Nationalist handling of the construction of Ping Shan airport, and the resumption of Chinese jurisdiction over the Kowloon Walled City.

---

3 TNA, CO 129/592/8, "extract from telegram No. 945 from Chongqing to Foreign Office", August 26, 1945.
4 TNA, FO 371/53639, "Memorandum of Agreements Relative to Japanese Surrender at Hong Kong", August 1945.
5 They were Nationalist crack troops and one army normally incorporated some 30,000 – 45,000 soldiers.
6 TNA, FO 371/53639, telegram from Mark Young (Hong Kong) to George Hall (CO), September 20, 1946.

## 1.1 The Construction of Ping Shan Airport

The British military government in Hong Kong (which was in place between September 1945 and April 1946) began to create a new defence plan for the colony at the end of 1945. A new airfield was deemed necessary. The existing Kai Tak airstrip on the northern shore of Kowloon bay, whose maximum runway length was only 1,400 yards, was not able to handle heavy aircraft, and its approaches were compromised by high hills, so it was not suitable for night flying. The new aerodrome planed for Ping Shan in the Northwest New Territories had a runway up to 3,000 yards and was suitable for use day or night.[7] A large area of land was, however, required to construct this new facility, which, according to British statistics, amounted to 600 acres and involved eight villages, 700 houses and 1,300 villagers.[8] But the local villagers were opposed to the project and asked for the Chinese government's intervention.[9] The Chinese press promptly linked Ping Shan with the unsettled future of Hong Kong. *Shishi Xinbao* 時事新報 in Chongqing stated that "Hong Kong and Kowloon should be negotiated to return to China."[10] Indeed, mass student demonstrations took place in Chongqing in January 1946 demanding the return of Hong Kong to China.[11]

The Nationalist authorities were obliged to intervene. A Chinese protest was sent through Guo Dehua 郭德華, the Chinese Special Commissioner for Foreign Affairs relating to Guangdong and Guangxi, to Admiral Harcourt, the Commander-in-Chief of Hong Kong.[12] The text reads as follows:

1. Kowloon though under lease to British, is Chinese territory and such a large-scale constructive project should not be undertaken without consent of [the] Chinese government.
2. Airfield construction causes displacement of [a] large number of Chinese nationals resident in that area whose interests are of concern to [the] Chinese government. At the time of conclusion of [a] new treaty with [the] British, [the] Chinese government reserved the question of Kowloon for negotiation after war.

---

7 TNA, FO 371/53629, "Hong Kong Defence Plan" annexed to telegram from Office of the Commander-in-Chief, Hong Kong, to the Secretary to the Chiefs of Staff Committee, London, December 24, 1945.
8 TNA, FO 371/53630, the document named "Ping Shan Airport", annexed to telegram from N.L. Mayle (CO) to Sterndale Bennett (FO), January 12, 1946.
9 TNA, FO 371/53629, telegram from Sir H. Seymour to Foreign Office, January 10, 1946.
10 Quoted from Sun, *Wuguoerzhong*, 185.
11 TNA, FO 371/53632, newspaper article clipped from *The Yorkshire Post* of January 28, 1946.
12 TNA, FO 371/53629, telegram (IZ. 52) from C-in-C, Hong Kong, to Cabinet Offices, January 3, 1946.

3. Pending a settlement of status of Kowloon the Hong Kong government is requested to discontinue the projected construction.[13]

The tone of the Chinese protest was strong, but it was not sent through regular diplomatic channels, such as the British Embassy in Chongqing or the Chinese Embassy in London. Guo's role, in British eyes, was tantamount to a Chinese Consul-General in Hong Kong who protected the interest of Chinese residents in Hong Kong. By using this lower-level channel, the Nationalist government was able to express its concerns but avoid triggering any formal discussions over Hong Kong's future.

This manoeuvre did not escape the notice of the British Ambassador, Horace Seymour. He inferred that "the whole agitation is insincere and is simply being used to keep the Hong Kong-Kowloon question alive."[14] Nevertheless, Sterndale Bennett and his impending successor, George V. Kitson (holding the position from 1946 to 1947), were concerned with the implications of the protest. It was argued by the Chinese that the New Territories, though under lease to Britain, were still Chinese territory and any large-scale construction projects required the consent of the Chinese government. In Kitson's view, such consent "would imply acceptance in advance of the Chinese claim for the rendition of the territories."[15] He and Sterndale Bennett were also confused as to why the Chinese had chosen to use this argument.[16] The 1898 convention seemed to give them more legitimate grounds for dispute:

> It is further understood that there will be no expropriation or expulsion of the inhabitants of the district included within the extension, and that if land is required for public offices, fortifications, or the like official purposes, it shall be bought at a fair price.[17]

A literal interpretation of this clause implied that the British could not compel a local person in this area to leave his land even if they gave him a fair price. Kitson was concerned that this more challenging line of argument would be pursued by the Chinese through "more formal representations".[18]

---

**13** The Chinese protest here is an original translation quoted from British sources. See TNA, FO 371/53629, telegram (IZ. 60) from C-in-C, Hong Kong, to Cabinet Offices, January 3, 1946.
**14** TNA, FO 371/53630, telegram from Sir H. Seymour to Foreign Office, February 1, 1946.
**15** TNA, FO 371/53630, minute by G.V. Kitson, January 12, 1946.
**16** TNA, FO 371/53630, minute by Sterndale Bennett, January 21, 1946.
**17** UK Treaties Online, "Convention between the United Kingdom and China respecting an Extension of Hong Kong Territory".
**18** TNA, FO 371/53630, minute by G.V. Kitson, January 12, 1946.

The Chinese government, however, was in no mood to inflame the dispute. Chiang Kai-shek hoped the agitation would quieten down if the British could be persuaded to abandon the airfield construction in Ping Shan,[19] and to this end he proposed opening the Baiyun airfield in Guangzhou to British air transport.[20] Meanwhile, the Chinese Foreign Minister telegraphed Guo to instruct him to engage with discussions with the British on the principle that the matter of land expropriation at Ping Shan should be separated from the case of Kowloon, which would be raised when international conditions were more advantageous to China.[21]

Further inquiries by a British Air Ministry Mission gave London an excuse to halt the Ping Shan plan. The Ping Shan southern approach was now considered by the Civil Aviation authorities as unsuitable for all weather service by heavy aircraft. An alternative site was thus proposed at Shenzhenwan, two miles north west of Ping Shan village.[22] In April 1946, the Ping Shan airfield project was formally abandoned.[23] Nevertheless, the Chinese outcry against the proposed airfield made the British more prudent about large-scale construction in the New Territories. For example, the building of a new naval transmitting and receiving station to support the British Pacific Fleet's operations was postponed at the suggestion of the Foreign Office,[24] although the New Territories were the only available option for the site.[25]

---

**19** AH Archives, 020 – 042701– 0027, *Ying zai Jiulong qiangzheng mindi jianzhu jichang* [British Compulsory Expropriation of Civil Land in Kowloon for Airfield Construction], telegram from Nationalist Government to the Foreign Ministry, January 11, 1946.

**20** AH Archives, 020 – 042701– 0027, *Ying zai Jiulong qiangzheng mindi jianzhu jichang* [British Compulsory Expropriation of Civil Land in Kowloon for Airfield Construction], letter from Zhou Zhirou 周至柔 to Wang Shijie, January 12, 1946.

**21** AH Archives, 020 – 042701– 0027, *Ying zai Jiulong qiangzheng mindi jianzhu jichang* [British Compulsory Expropriation of Civil Land in Kowloon for Airfield Construction], telegram from Foreign Ministry to Guo Dehua, March 28, 1946.

**22** TNA, FO 371/53632, telegram from Commander in Chief, Hong Kong to Cabinet Offices, February 15, 1946.

**23** TNA, FO 371/53634, letter from Colonial Office to Mr. Cribbett (Civil Aviation), April 27, 1946. The construction of the airfield in Shenzhenwan was not carried out in the end due to the fact that the location was very near to the China-Hong Kong boundary. The Hong Kong government then expanded the old Kai Tak aerodrome instead in the 1950s.

**24** TNA, FO 371/53629, letter from G.V. Kitson (FO) to R.R. Powell (Admiralty), January 25, 1946.

**25** TNA, FO 371/53629, letter from R.R. Powell (Admiralty) to J.J. Paskin (CO), January 8, 1946.

## 1.2 The Kowloon Walled City

In contrast to the cautious policy of the Nationalist authorities over the airfield, ardent advocates for Hong Kong's return in the Guomindang argued for more decisive action in relation to the Kowloon Walled City. Lin Xiazi 林俠子 was a prominent example. Lin, the magistrate of Bao'an county, which was adjacent to Hong Kong, and his main aide Tan Jialin 譚家琳, head of the Civil Affairs Department in the county, were eager to resume Chinese civil administration over the city. The city, as well as the remainder of the New Territories, had been part of Bao'an county in the pre-cession era. After examining the 1898 convention, Lin was convinced that it should still be under Chinese sovereignty.[26] It was stipulated in the convention that

> within the city of Kowloon, the Chinese officials now stationed there shall continue to exercise jurisdiction except so far as may be inconsistent with the military requirements for the defence of Hong Kong. Within the remainder of the newly-leased territory Great Britain shall have sole jurisdiction.[27]

For this reason, the Chinese Foreign Ministry gave its tacit support to Lin's effort and assigned Guo Dehua to assist him diplomatically.[28]

However, the matter was leaked to the public by the Chinese press before any practical plans had been put in place.[29] The Hong Kong government immediately took a strong stance. An official statement was issued on September 15, 1946, before any consultation with London, denying that the Kowloon Walled City remained under Chinese jurisdiction:

> In 1899 the exercise of jurisdiction by Chinese officials in Kowloon City [the Kowloon Walled City] was found to be inconsistent with the military requirements for the defence of Hong Kong and was accordingly terminated. Since that date, that is for nearly 50 years,

---

**26** TNA, CO 537/1657, minutes of the talk between Mark Young and Guo Dehua, recorded by Mark Young, September 16, 1946.

**27** UK Treaties Online, "Convention between the United Kingdom and China respecting an Extension of Hong Kong Territory".

**28** IMH Archives, 11–32–27–01–003, *Jiulong chengzhi jiufen* [Disputes over the Jurisdiction of the Kowloon City], telegram from Ministry of Foreign Affairs to Guo Dehua, August 10, 1946.

**29** TNA, CO 537/1657, extract from the "HSIN SHENG WAN PAO" [*Xinsheng Wanbao* 新生晚報] (dated August 6, 1946) and minutes of the talk between Mark Young and Guo Dehua, recorded by Mark Young, September 16, 1946.

the British authorities have exercised sole jurisdiction in Kowloon City as in the remainder of the New Territories. [30]

Such an attitude caused a sharp reaction from the Chinese authorities. The next day, Guo Dehua met Mark Young, Hong Kong's Governor. The former indicated that the 1899 order was a "unilateral action" from the British side which the Chinese government had never recognised. [31] Three further points were made:

a. The Chinese inhabitants of Kowloon City [the Kowloon Walled City] had at various times since 1899 paid taxes to the Chinese Authorities;
b. The interior of Kowloon City [the Kowloon Walled City] had been entirely neglected by the Hong Kong government, the roads were in poor condition, and the police never entered there;
c. Some sort of negotiations with the Hong Kong government had been conducted by his predecessors [in the 1930s]. [32]

Young, however, refused to discuss the setting up of a Chinese civil administration in the Kowloon Walled City. [33]

Yet, as a consequence of the recalcitrant attitude of the statement of the Hong Kong government, numerous protests from Chinese authorities at province, city and county levels, particularly the Guangdong provincial government, urged the central government to take a tough stand against the British, even urging it to take the opportunity to retrieve the whole of Hong Kong. [34] The Press lobbied hard. An editorial in the *Zhongyang Ribao* of September 19 argued that

For future good relations between China and Britain and in the interest of their joint efforts to promote world peace, Britain must restore Hong Kong and Kowloon to China, ... The first move in this direction should be the abolition of the Treaty of Peking [Beijing] of 1898. China's sovereignty of Kowloon City [the Kowloon Walled City] should not be declared null and void by the Hong Kong government on the pretext of running counter to British military requirements. This Sino-British Treaty of Peking was the result of British colonial policy in the Far East, and the Kowloon problem arose out of the policy of the British colonial authorities in Hong Kong. The Treaty of Peking is one of the unequal treaties forced on China by Britain

---

**30** TNA, CO 537/1657, "Official Statement Issued by Hong Kong Government", September 15, 1946.
**31** TNA, CO 537/1657, minutes of the talk between Mark Young and Guo Dehua, recorded by Mark Young, September 16, 1946.
**32** Ibid.
**33** Ibid.
**34** IMH Archives, 11–32–27–01–003, *Jiulong chengzhi jiufen* [Disputes over the Jurisdiction of the Kowloon City], letters from Guangzhou Provisional Consultative Council to Ministry of Foreign Affairs, October 4, 1946.

and the abolition of the treaty should be the key to lasting understanding between [the] two countries. The Chinese Government's policy is that any change in the status of the city of Kowloon should be brought about through diplomatic negotiations and neither China nor Britain should take unilateral action.[35]

In the Foreign Ministry Chinese officials laid out three possible ways to handle the issue:

a. To affirm Chinese treaty rights concerning the Kowloon city [the Kowloon Walled City] while making an understanding with the British. That being so, the Chinese would temperately refrain from establishing civil governance over the city but reserved the rights to do so at any time.

b. To inform the British of [Chinese treaty rights over the city] and meanwhile proceed with coordination between the Ministry of the Interior and the Ministry of Justice in establishing civil institutions and a legal court in the city. Once the preparation was completed, a Chinese civil administration and legal system must be set up no matter with British concurrence or not.

c. To recover the whole Kowloon leased areas [the New Territories] by this opportunity.[36]

The Foreign Ministry preferred the first option since it was considered difficult for the Chinese to govern the city without British cooperation as it was surrounded by British-administrated areas.[37] The Ministry also had no wish for a showdown with the British. Setting up administrative institutes was deferred until a more suitable time emerged.[38]

The British authorities equally wanted to defuse any tensions emanating from the Kowloon Walled City. Ralph Stevenson, the new British Ambassador at Nanjing, thought that the Chinese Ministry of Foreign Affairs had "allowed themselves to be put into a false position by [a] party of hot-heads", and that its recent intervention seemed to "show disinclination to allow the [launch] of a new campaign about Hong Kong."[39] In this regard, he was inclined to "[wait] to see whether the press agitation, which has died down, is renewed."[40] In line with this, the Foreign Office advised Stevenson to pursue "a completely

---

**35** TNA, CO 537/1657, extract from information released by UK office of the Chinese message of service on the question of Hong Kong and Kowloon, October 2, 1946.

**36** IMH Archives, 11–32–27–01–003, *Jiulong chengzhi jiufen* [Disputes over the Jurisdiction of the Kowloon City], minute by the Ministry of Foreign Affairs, October 2, 1946.

**37** Ibid.

**38** IMH Archives, 11–32–27–01–003, *Jiulong chengzhi jiufen* [Disputes over the Jurisdiction of the Kowloon City], telegram from Ministry of Foreign Affairs to Guo Dehua, August 9, 1947.

**39** TNA, CO 537/1657, telegram from Sir R. Stevenson to Foreign Office, September 24, 1946.

**40** Ibid.

non-committal attitude" in reply to any questions about Hong Kong and the New Territories.[41] It also suggested that the Governor of Hong Kong should take "an exactly similar line" but that such messages should be delivered by the Colonial Office.[42] There was some hesitancy about how to draw the Governor's attention to this protocol and avoid fettering the latter unduly.[43] As a result, the following diplomatically worded telegram was sent:

> the possible repercussion outside Hong Kong, even of action intended only to pacify local opinion, may sometimes be considerable. For that reason, it is generally advantageous to consider the question of an approach through diplomatic channels before any other action is taken. ... [You] will consult the Embassy and [the Colonial Office] in advance about any proposals, particularly when they involve publicity, for countering future Chinese propaganda or other activities affecting the status of Hong Kong or the New Territories.[44]

In the end the dispute over the Kowloon Walled City was once again shelved.

## 1.3 Other Inflammatory Incidents

The recovery of Hong Kong remained a pressing issue for Chinese public opinion as a result of further inflammatory incidents. On October 26, 1946, Wang Shuixiang 王水祥 (Wang Shui-hsiang), a Chinese peanut hawker, was kicked to death in Kowloon by a Portuguese member of the Hong Kong police force.[45] This crime occurred during a campaign by the Hong Kong police to clear hawkers from the streets in Kowloon and Hong Kong, and over the previous week hundreds of such hawkers had been arrested and their goods confiscated.[46] This prompted a demonstration by 1,000 Chinese protestors. The police opened fire over their heads to disperse the crowd but injured many.[47] The same mass indignation was on display on the day of the hawker's funeral.[48] These protests, as well as

---

41 TNA, CO 537/1657, telegram from Foreign Office to Sir R. Stevenson, September 26, 1946.
42 TNA, CO 537/1657, minute by N.L. Mayle (CO), September 25, 1946.
43 TNA, CO 537/1657, minute from T.K. Lloyd (CO) to Sir G. Gater (CO), October 1, 1946.
44 TNA, CO 537/1657, draft telegram from Colonial Office to Hong Kong, undated, and minute by N.L. Mayle (CO), September 30, 1946.
45 TNA, FO 371/63386, memorandum from Ministry of Foreign Affairs to HBM Embassy, December 7, 1946.
46 TNA, FO 371/53638, "Chinese Hawker Killed by Portuguese Policeman", extract from news item issued by London Office of the Chinese Ministry of information, November 1, 1946.
47 TNA, FO 371/53638, "Chinese Demonstration in Kowloon", extract from news item issued by London Office of the Chinese Ministry of information, November 1, 1946.
48 TNA, FO 371/53639, telegram from Hong Kong to Colonial Office, November 15, 1946.

a Chinese press campaign against the Hong Kong government, were directed by the Guangzhou branch of the Guomindang.[49] The Guangdong People's Political Council also passed a resolution to impose a boycott upon trade with Hong Kong.

However, the Chinese central government wished for no boycotts or any other actions likely to disturb Sino-British relations.[50] Wang Shijie, the Chinese Foreign Minister, hoped that the case would be handled expeditiously and explained to Stevenson that the Chinese Foreign Ministry had always done its best to prevent any exacerbation of tension over such matters but was under continual criticism from Guangzhou for being weak-kneed: the Guomindang organisation was rather loose-knit and its policy was not necessarily that of Central Committee of the Guomindang.[51]

Yet another tragic incident kept the issue simmering. On December 3, 1946 a Chinese villager, Zhang Tianxiang 張添祥 (Chang Tien-hsiang), was shot to death by a British soldier when the former was passing through the Hong Kong-China border. About a hundred Chinese villagers held a demonstration two days later at Shenzhen 深圳, adjacent to Hong Kong, protesting against this border incident. The villagers also organised a meeting at which they requested the Chinese government to take immediate steps to negotiate the return of Kowloon to China.[52]

Meanwhile, numerous letters of protest from the consultative councils of nearly every country, city and province in China flocked into the central government in Nanjing during 1946 and 1947, urging the latter to negotiate the recovery of Hong Kong and Macao.[53] The Guomindang authorities were united in their desire for the eventual recovery of Hong Kong and Macao to China, but opinions were divided over the timing and approaches to achieving this. Most local authorities leaned towards immediate action. The Foreign Ministry, however, wish-

---

**49** TNA, FO 371/53639, telegram from Sir R. Stevenson (Nanjing) to Hong Kong, December 7, 1946.

**50** TNA, FO 371/53639, telegram from Sir R. Stevenson (Nanjing) to Hong Kong, November 30, 1946.

**51** TNA, FO 371/53639, telegram from Sir R. Stevenson (Nanjing) to Hong Kong, December 7, 1946.

**52** TNA, FO 371/53639, "New Messages from [Nanjing]", December 6, 1946.

**53** Four archive files are filled with these protesting letters: IMH Archives, 11–01–19–04–03–012, *Shouhui Gang Ao ji Jiulong gefang yijian; shouhui Gang Ao wenti* [Various Opinions over the Recovery of Hong Kong, Macao and Kowloon: the Question of Restoring Hong Kong and Macao] and AH Archives, 001–062001–0008, 001–062001–0009, 001–062001–0010, *Zhong-Ying Gao Ao jiufen jiaoshe 1, 2 and 3* [Negotiations about Conflicts Arising from Hong Kong and Macao Volumes 1, 2 and 3].

ed to postpone negotiations until the international and domestic situations were more suitable. In September 1946, it maintained that

> The three territories are distinctive in nature: Hong Kong is the ceded land; Kowloon is the leased land; and Macao was a special residence settlement. Jurisdiction over the Kowloon Walled City has never been abandoned by any Chinese regimes and, therefore, the scheme of the local government to resuscitate its governance over the city can be continued. As far as the questions of Hong Kong and Macao are concerned, recovery of these two territories is the Nationalist government's set policy and for which the Ministry [of Foreign Affairs] has already started preparing specific formulas. The difficulties lie in the timing. Serious consideration must be made to determine whether the present time is appropriate since the current situations are fraught with tremendous troubles. ... It is advisable to defer negotiations about Hong Kong, Kowloon, and Macao until peace and unification return to China and until international conditions are advantageous to the government. The Chinese government will not take any active approaches at the present time.[54]

With reference to the means, Chen Anren 陳安仁, a member of the Foreign Affairs Committee of the Legislative Yuan, suggested the government retake Hong Kong by instalments, negotiating the leased territories first and then the ceded land, and at the same time Chinese jurisdiction could be restored over the Kowloon Walled City.[55] Yet the Foreign Ministry decided to consider possible tactics at a later stage.[56]

Nevertheless, the Nationalists continued to test the British resolve. In June 1946, Chiang Kai-shek told Seymour, shortly before his retirement and return to Britain, that amicable relations between China and Britain would be difficult to achieve unless the Hong Kong problem was solved satisfactorily.[57] In London, Gu Weijun, the Chinese ambassador to London, took the opportunity, when bidding farewell to the Foreign Office, to seek out the British position over Hong Kong. Philip Noel-Baker, the Minister of State, informed him that it was not the time to discuss the Hong Kong issue while Britain was preoccupied with

---

**54** IMH Archives, 11–01–19–04–03–012, *Shouhui Gang Ao ji Jiulong gefang yijian; shouhui Gang Ao wenti* [Various Opinions over the Recovery of Hong Kong, Macao and Kowloon: the Question of Restoring Hong Kong and Macao], minute by the Ministry of Foreign Affairs, September 27, 1946.

**55** IMH Archives, 11–01–19–04–03–012, *Shouhui Gang Ao ji Jiulong gefang yijian; shouhui Gang Ao wenti* [Various Opinions over the Recovery of Hong Kong, Macao and Kowloon; The Question of Restoring Hong Kong and Macao], proposal of the Legislative Yuan, October 2, 1946.

**56** IMH Archives, 11–01–19–04–03–012, *Shouhui Gang Ao ji Jiulong gefang yijian; shouhui Gang Ao wenti* [Various Opinions over the Recovery of Hong Kong, Macao and Kowloon; The Question of Restoring Hong Kong and Macao], letter from the Ministry of Foreign Affairs to the Committee of Foreign Affairs, the Legislative Yuan, October 9, 1946.

**57** TNA, FO 371/53635, Annex 1 to "Future of Hong Kong", July 18, 1946.

problems in India and Egypt.[58] Despite this, the Chinese refused an informal British proposal to base the United Nations' headquarters in the Far East and Pacific region[59] at Hong Kong in case this further complicated the status of Hong Kong.[60]

## 2 British Policy Options regarding Hong Kong's Future, 1945–1947

The transfer of power from the Conservative Party to the Labour Party in the summer of 1945 did little change to Britain's strategy towards China and Hong Kong.[61] The future of Hong Kong was still a difficult matter for the British to wrestle with but, as Sterndale Bennett put, "has got to be squarely faced."[62] In a memorandum of January 1945, entitled "The Status of Hong Kong", the difficulties that the British faced had been summarised. This paper had been formulated by the Far Eastern Civil Planning Unit, an arm of the Far Eastern Committee, an inter-departmental committee tasked with considering Britain's policy towards East Asia. It contended that Hong Kong was very important to British interests, whether economic, emotional, or strategic. In view of the absence of extraterritorial protection, and the anticipated unstable political and economic situation in China for some years to come, Hong Kong would be one of a few safe bases from which British merchants and industrialists could operate within China. The British public also had some sentimental attachment to Hong Kong. The colony had been "a desolate island" when it was ceded to Britain in 1842 and it was widely believed that it was "British enterprise, law and security" that had built it up to be one of great seaports of the world. Furthermore, Hong Kong formed part of a forward defensive system protecting the main British strategic

---

58 IMH Archives, 11–01–19–04–03–012, *Shouhui Gang Ao ji Jiulong gefang yijian; shouhui Gang Ao wenti* [Various Opinions over the Recovery of Hong Kong, Macao and Kowloon: the Question of Restoring Hong Kong and Macao], telegram from Gu Weijun to the Ministry of Foreign Affairs, June 29, 1946.

59 The establishment of the UN Office in the Far East and Pacific region was a proposal at the time in response to the setting of the UN Office in Europe and Central Asia at Geneva but did not materialise. Only a few UN regional originations were based in Shanghai before 1949, such as the United Nations Economic Commission for Asia and the Far East (UNECAFE).

60 IMH Archives, 11–01–19–04–03–012, *Shouhui Gang Ao ji Jiulong gefang yijian; shouhui Gang Ao wenti* [Various Opinions over the Recovery of Hong Kong, Macao and Kowloon: the Question of Restoring Hong Kong and Macao], letter from the Ministry of Foreign Affairs to Chang Kai-shek, November 15, 1946.

61 Ovendale, "Introduction," 1–12.

62 TNA, FO 371/53632, minute by Sir John C. Sterndale Bennett, March 2, 1946.

areas of interest in East and Southeast Asia. Any British concessions in Hong Kong might well be regarded as a sign of "weakness" and "stimulate agitation in regard to Chinese claims and ambitions in South East Asia."

However, these benefits, as well as Hong Kong's own prosperity, to a great extent, relied on a harmonious Sino-British relationship. As the paper pointed out, "a hostile Chinese government could probably make life impossible for [the British] in Hong Kong, even if it was not in possession of the New Territories: this is indicated by the strike and boycott of 1925–26 which virtually paralysed the trade of the port."[63] To solve the dilemma, the memorandum suggested four possible solutions. The first was that Britain should not yield to any Chinese claims. The second was the retrocession of the New Territories with certain conditions attached:

a. An Anglo-Chinese joint board of management for the airfield and for the water storage and supply system;
b. A Joint Municipal Board [of management] for the urban district of the New Territories adjoining the British portion of Kowloon;
c. Chinese government representation on a Port Authority for Hong Kong.[64]

The third possibility was British cession of its sovereignty over Hong Kong in exchange for a lease or other such arrangement.[65] The fourth was to return all territories to China. This was a last resort and could only be contemplated if:

a. It is evident that a strong and just government has been established in China which has shown itself able and willing to afford security and fair-trading conditions for foreign enterprise in China;
b. The strategic implications of new weapons such as rocket projectiles and atomic bombs have been fully worked out.[66]

The memorandum was classified as "top secret" and had been circulated to other departments for comment. Yet it remained purely a discussion document.

---

**63** TNA, CO 537/1656, a copy of an annex to a document entitled "British Policy in the Far East", affixed to a Colonial Office's minute, April 16, 1946.
**64** Ibid. The only airport in Hong Kong at this time, the Kai Tak airfield, and the main source of its fresh water, the Shing Mun reservoir, were both in the New Territories.
**65** Ibid.
**66** Ibid.

## 2.1 The Foreign Office's Memorandum on Hong Kong, 1946

Chinese protests and boycotts against the British galvanised the Foreign Office into re-evaluating Hong Kong's future. Triggered by the disputes over Ping Shan airport, Kitson started to prepare a fresh memorandum in February 1946, on the basis of the aforementioned 1945 memorandum, entitled "The Future of Hong Kong". He suggested that the British government take the initiative and ne- gotiate about the return of Hong Kong to China.[67]

Kitson's proposal was understandably controversial. Sterndale Bennett doubted whether there was "any really practicable half-way house",[68] while Har- old Orme Garton Sargent, the Permanent Under-Secretary for Foreign Affairs, contradicted the view that "concessions to China in regard to Hong Kong are ei- ther necessary or desirable in order to buy Chinese goodwill."[69] Mark Young and Horace Seymour were also consulted.[70] They too doubted whether the Chinese would be satisfied with limited British concession over Hong Kong. Seymour was convinced that "in any event the usual Chinese technique would be to take what they could get as a first instalment towards the attainment of their full object."[71]

The memorandum was reworked, and another version produced in July 1946. It was recognised that the Hong Kong problem could not be handled in iso- lation, or in disregard of the colony's relations with China, and that the close ra- cial, geographical and commercial links between the colony and the mainland should not be understated. Furthermore, as had been demonstrated during the boycotts following the May Thirtieth Incident in 1925–1926, China had the ca- pacity to paralyse the British administration of Hong Kong. And even if the Chi- nese government took no part in any direct action, the danger always existed of local Chinese extremists getting out of hand and undermining the British posi- tion in Hong Kong by use of strikes and boycotts. In this regard, it was thus sug- gested that the colony might conceivably be returned to China but only on the basis that Britain was granted "a thirty-year lease of the whole areas" as well as a commitment that Hong Kong would continue to function beyond that as a secure commercial base for British interests. To strengthen the case for such a lease, it was advised that during the lease term "Hong Kong, in addition to being a free port, would be available for the joint use of Britain and China as

---

**67** TNA, FO 371/53632, memorandum by G.V. Kitson, February 28, 1946.
**68** TNA, FO 371/53632, minute by Sir John C. Sterndale Bennett, March 2, 1946.
**69** TNA, FO 371/53632, minute by Sir Harold Orme Garton Sargent, March 19, 1946.
**70** Seymour left his position as British Ambassador to China in May 1946.
**71** TNA, CO 537/3712, "Consideration of Development Policy in New Territories", October 7, 1948.

a naval and air base for maintaining security in the Pacific." To make the prospect more palatable to the Chinese, the paper also suggested a customs agreement since smuggling from Hong Kong to mainland China had long been a thorn in China's side.[72]

## 2.2 The Colonial Office's Memorandum on Hong Kong, 1947

For their part, the Colonial Office and the Hong Kong government were agitated by frequent Chinese interventions in local Hong Kong issues. In their eyes, this undermined the allegiance of the residents to British authority and made British administration in the colony difficult. It also raised concerns about Hong Kong's longer-term economic development and future investments in the New Territories. At the end of 1946 the Colonial Office contemplated a joint Colonial Office-Foreign Office memorandum on Hong Kong. As part of this, it wanted a public reassurance that the British government would not abandon Hong Kong and clarification over what the bottom line would be if the Chinese started formal negotiations over the colony. In its view, Britain could only return the leased portion of Hong Kong to China on certain conditions.[73] These were as follows:

a.  The re-drawing of the frontier between ceded territories and the leased territories so as to include with the ceded territory the whole of the built-up area of Kowloon and certain islands which at present fall within the leased territory. ...

b.  Joint management of Kai Tak airfield and of that part of the water supply system which would run beyond the Colony's new boundary. ... [An] understanding about the grant of facilities to the Hong Kong government for constructing any further installations in the New Territories (e.g. a new airfield) which might be regarded as necessary or desirable in the interests of the Colony.[74]

The Foreign Office, however, was sceptical. In particular, it objected to any public statement of reassurance over the retention of Hong Kong.[75] Such a statement, from its perspective, might cause the Chinese to feel impelled to increase their pressure, either provoking premature Anglo-Chinese negotiations or sparking

---

72 TNA, FO 371/53635, memorandum of 'The Future of Hong Kong' by the Foreign Office, July 18, 1946.
73 TNA, FO 371/63386, "The Future of Hong Kong", revised draft of Joint Memorandum by the Foreign Office and the Colonial Office, attached to letter from N.L. Mayle (CO) to G.V. Kitson (FO), January 14, 1947.
74 TNA, CO 537/2190, "Conditions, suggested by the Governor of Hong Kong, to be attached to the surrender of the lease of the New Territories" by the Colonial Office, October 1946.
75 TNA, FO 371/63386, minute by G.V. Kitson, January 20, 1947.

Chinese riots against British interests.[76] A joint memorandum was thus never agreed.[77] Nevertheless, it now seemed to be evident that the Colonial Office and the Foreign Office were broadly in agreement that some sort of compromise over Hong Kong was inevitable.

## 3 The Hong Kong Residents' Attitudes toward the Reversion of Hong Kong to China

In spite of the fact that Hong Kong's future was to be decided by the British and Chinese governments, the attitude of Hong Kong residents was still a factor. There had been big demographic changes in the 1940s. During the Japanese occupation of Hong Kong from 1942 to 1945, some 554,000 people had been repatriated to China with the purpose of lessening Japan's problems in maintaining law and order, housing, food and local defences.[78] The population of Hong Kong had shrunk to only 600,000 by the war's end.[79] From 1945 to 1948 many returned and a large number of new Chinese immigrants arrived, swelling the population to approximately 1.6 million.[80] The civil war in China gave rise to a large further intake of refugees to Hong Kong, especially from South China. The population of the colony soared to 2,360,000 by 1950.[81] The main element of the Hong Kong population was thus first-generation Chinese immigrants, who were either looking for economic opportunities or took refuge to escape political or military conflict. As regards their allegiance to British rule in Hong Kong, the following observations provide some useful insights.

Chen Yifan 陳依範 (Jack Chen), who was the son of Chen Youren 陳友仁 (Eugene Chen), a former Chinese foreign minister, and lived in Britain, commented after a four-month visit to China and Hong Kong in 1947 that

> No progressive group of Chinese in Hong Kong or China wants Hong Kong to be handed over to the [Guomindang] as it is at present constituted. But every Chinese I have met wants Hong Kong to be returned to China at some stage when a united and democratic Chi-

**76** TNA, FO 371/63386, letter from Foreign Office to A. Creech-Jones, January 28, 1947.

**77** TNA, CO 537/3712, "Consideration of Development Policy in New Territories", October 7, 1948.

**78** Wing-Tak Han, "Bureaucracy and the Japanese Occupation of Hong Kong," in *Japan in Asia, 1942–1945*, ed. William H. Newell (Singapore: Singapore University Press, 1981), 14.

**79** Chi-Kwan Mark, "The 'Problem of People': British Colonials, Cold War Powers, and the Chinese Refugees in Hong Kong, 1949–62," *Modern Asian Studies* 41, no. 6 (2007): 1146.

**80** John P. Burns, "Immigration from China and the Future of Hong Kong," *Asian Survey* 27, no. 6 (1987): 662.

**81** Mark, "The 'Problem of People'", 1146.

nese government can guarantee to the people of Hong Kong the same level of well-being and hope of progress that they now enjoy. In Chinese breasts, the tide of nationalism runs strong. It will not be stemmed.[82]

The *Times* reported similar sentiments on June 2, 1948:

Most of the Chinese inhabitants have no desire to see the chaos prevailing at Shanghai introduced into Hong Kong. ... These Chinese are not pro-British, but they want a continuance of British administration. Their trouble is that they are frightened; they want to be in with both sides. It is easy for [the Guomindang] to exert pressure on them directly and indirectly, as nearly all have interests and relatives in China. This fear has been accentuated by the apparent indecision of His Majesty's Government since the end of the war.[83]

Duncan J. Sloss provided a further view. He was the Vice-Chancellor of University of Hong Kong (1937–1949) and a former member of the Governor's Executive Council (1940–1941). He had lived in Hong Kong for many years and felt despondent about the prevailing anti-British views among Chinese communities in Hong Kong. He attributed these to Britain's short-sighted policy, noting that over the previous hundred years, Hong Kong had been treated as "the shop window of Great Britain in China" and there was never enough money for education, public health and social services before the war "though Hong Kong was probably the most lightly-taxed community in the Empire." The government had done "almost nothing by education, social services or political education to foster a "Hong Kong" patriotism among the Chinese."[84] Only two groups in his eyes were "reliable". The first was the Hong Kong-born Chinese, "who appreciated the value of a stable government." This group included many who had relatives in "Malaya, Canada and Australia as well as in China proper and also among the Hong Kong Eurasian community." The second group was centred on the Democratic League. It contrasted the liberal views of the West with those of the Chinese government.[85] Yet these two groups were apparently in a minority in Hong Kong in the 1940s. All in all, it could be concluded that most Hong Kong residents appreciated the stability of Hong Kong under British administration but did not necessarily identify themselves with either the colony or with Britain.

---

**82** TNA, FO 371/63390, article of "A New Hong Kong" by Mr Jack Chen, July 1947.
**83** TNA, FO 371/69582A, article of "Pressure Upon Hong Kong: Divided Loyalty of the Chinese Population" slipped from *The Times* of June 2, 1948.
**84** TNA, FO 371/53638, letter from D.J. Sloss to Sir Stafford Cripps (BT), copied to Foreign Office, October 11, 1946.
**85** Ibid.

## 4 The Shifting Focus of the Nationalist Government Position on Hong Kong, 1947–1948

The Guomindang's focus on the recovery of Hong Kong shifted again in 1947–1948. At this juncture, the Nationalists were more concerned with widespread smuggling and Chinese monetary speculation in the colony. In Wang Shijie's eyes, Hong Kong was becoming "a political and economic menace to China."[86] To be fair, the economic and financial difficulties that the Nationalist government faced at this time were principally caused by its "deficit financing" policy, which had been frequently used between 1937 and 1945.[87] In 1947–1948, the gap between the government's revenue and expenditure was speedily expanded by escalating hostilities between the Guomindang and the CCP in Northeast China and North China. Nationalist authorities were obliged to rely on the printing press as the principal means for wiping out the deficit. This generated rampant hyperinflation in Nationalist-controlled areas.[88] Meanwhile, to increase the government's revenue collection, heavy taxes were levied on trade and commodities, and more control was imposed on exports, imports and currency exchange. The high customs tariffs and a ban on imports considered "inessential" prompted tremendous smuggling. An overvalued official exchange rate for the Chinese dollar also drove merchants to export goods via smuggling so that they could exchange their earned US dollars or sterling for Chinese dollars at a much higher price on the black market. Against such a background, Hong Kong, a traditional entrepôt for South China with an open exchange market, naturally became a haven for smuggling and Chinese monetary speculation.[89] In this context, the Nationalist government was eager for the Hong Kong government's cooperation in curtailing these activities.[90]

The Colonial Office hoped to seize this opportunity to secure Chinese assent for British control over Hong Kong.[91] The Hong Kong government was of the same mind. Mark Young noted that

---

**86** TNA, FO 371/69582A, telegram from British Embassy (Nanjing) to Foreign Office, May 31, 1948.

**87** Boecking, *No Great Wall*, 230–231.

**88** Suzanne Pepper, *Civil War in China: the Political Struggle, 1945–1949* (Lanham, Md.; Oxford: Roman & Littlefield Publishers, 1999), 95.

**89** Catherine R. Schenk, "Commercial Rivalry between Shanghai and Hong Kong during the Collapse of the Nationalist Regime in China, 1945–1949," *The International History Review* 20, no. 1 (1998): 76–77.

**90** TNA, FO 371/53637, telegram from Sir R. Stevenson to Hong Kong, August 16, 1946.

**91** TNA, T 236/681, aide-memoire, no date [late February 1947]. Quoted in Schenk, "Commercial Rivalry between Shanghai and Hong Kong, 1945–1949," 79.

In these circumstances it seemed most necessary that every effort should be made to secure a measure of agreement sufficient to satisfy reasonable Chinese expectations while at the same time surrendering as few as possible of the advantages which Hong Kong now enjoys. ... [The] comparatively small loss to the sterling area must be weighed against the ill-feeling and renewed agitation for the return of Hong Kong which will most certainly follow if no settlement is reached.[92]

In October 1947, Alexander Grantham, the new Governor of Hong Kong, paid an official visit to Nanjing. He was warmly welcomed by the Nationalist regime. Chiang Kai-shek, Madame Chiang Kai-shek, Zhang Qun 張群, the new president of the Executive Yuan, Sun Ke 孫科, the Vice-President, and other high-level Guomindang officials all entertained him. In the subsequent conversations, the question of Hong Kong's return was overridden by the urgent Nationalist desire for close and cordial cooperation with the Hong Kong government over the ending of smuggling and black-market dealing.[93]

Chinese priorities manifested themselves in two bilateral agreements. A customs agreement (signed on January 12, 1948) allowed Chinese customs officials to be stationed in Hong Kong to patrol the waters between Hong Kong and Guangdong, while a financial agreement (signed on August 15, 1948) provided that Hong Kong should share its accrual of US dollars according to a formula of returning them to China if they were earned through Chinese exports via the colony. [94] By way of these treaties, the Nationalist authorities aimed to restrict Chinese imports and exports through Hong Kong and drive these trades back to Shanghai and other Chinese ports under their direct control. They also intended to control foreign currency movements between Hong Kong and China and stem the large illegal outflow of capital away from China through Hong Kong.[95] However, from Hong Kong's perspective, the two agreements carried a risk of destroying the transit trade which dominated the colony's economy. The head of Hong Kong's department of trade, industry and supplies complained that "it is fairly obvious that for political reasons we are about to commit economic suicide, for there is now no money available from any source to finance the entrepôt trade on which we depend for our complete existence."[96] Therefore the two agreements were never strictly implemented by the Hong Kong govern-

---

**92** TNA, T 236/681, telegram from Hong Kong to Colonial Office, February 24, 1947. Quoted in ibid., 80.
**93** TNA, FO 371/63391, telegram from Sir R. Stevenson to Foreign Office, October 6, 1947.
**94** Schenk, "Commercial Rivalry between Shanghai and Hong Kong, 1945–1949," 79; ibid., *Hong Kong as an International Financial Centre*, 22–23.
**95** Ibid., "Commercial Rivalry between Shanghai and Hong Kong, 1945–1949," 81.
**96** TNA, OV 104/87, memorandum by Thomson, July 21, 1947. Quoted in ibid., 82.

ment.[97] In July 1948 a formal Chinese complaint was made to the Hong Kong government about the delay in enactment.[98]

To compound matters, in 1948 the Chinese economy began to disintegrate under the pressure of political upheaval and spiralling inflation. Business enterprises escaped to Hong Kong from Shanghai and other parts of China, generating substantial influx of capital. The Chinese public were also unwilling to entrust their savings to banks, and gold, silver and remittances also flowed into Hong Kong. The Bank of England estimated in April 1948 that about US$20 million per month was being deposited in US dollar accounts in Hong Kong.[99] The Guomindang's monetary reforms in mid-1948, coercing the Chinese to hand their gold, silver and foreign currency to the government in exchange for a Chinese new paper currency, the Gold Yuan, exacerbated the problem. The new currency was introduced at the rate of 1: 3,000,000 against the old Chinese dollar. The Nationalist government wished to use this as a hedge against inflation but, in fact, it was seen as a means to exploit the masses' wealth before it fled to Taiwan.[100] In such a situation, the Nationalist government was in no mood to press for the reversion of Hong Kong.

Alongside these economic concerns, the relatively liberal environment in Hong Kong became a haven for various anti-Nationalist elements. The colony was condemned by the Nationalist regime in 1948 as "a hot bed of anti-government intrigues and a centre of seditious propaganda designed to undermine the prestige of the Chinese government."[101] Indeed, the CCP used the colony as a base to facilitate its war with the Guomindang in South China. In January 1947, a Hong Kong Central Bureau was established in the colony with the purpose of organising propaganda campaigns against Chiang Kai-shek and the United States, which was regarded as Chiang's main supporter. Soon the Bureau developed into a headquarters for the CCP's underground activities in six provinces: Guangdong, Guangxi, Fujian, Hunan, Yunnan and Guizhou.[102] The CCP believed that its efforts should be concentrated on mainland issues and that it should avoid provoking the British authorities in Hong Kong. As one Brit-

---

**97** Ibid., 83–87.

**98** TNA, FO 371/69582A, aide-memoire by the Chinese Embassy at London, July 21, 1948.

**99** TNA, T 236/2033, report by Portsmore, the Bank of England representative on a visit to Hong Kong, April 20, 1948. Quoted in Schenk, "Commercial Rivalry between Shanghai and Hong Kong, 1945–1949," 86.

**100** Pepper, *Civil War in China: the Political Struggle, 1945–1949*, 121–131.

**101** TNA, FO 371/69582A, aide-memoire by the Chinese embassy in London, July 21, 1948.

**102** Christine Loh, *Underground Front: the Chinese Communist Party in Hong Kong* (Hong Kong University Press, 2010), 70–71.

ish official commented, "Hong Kong as a base from which to direct activities abroad is too valuable to be frittered away in local activities which might result in the banning or even expulsion of the [Communist] Party with its leaders."[103] Furthermore, Hong Kong was a good place for the CCP to establish contacts with other anti-Nationalist groups exiled in the colony, such as the China Democratic League (CDL) and the Guomindang Revolutionary Committee (GRC), so as to build a broad alliance against the Guomindang. The former, established mainly by intellectuals, was proscribed by the Nationalist government, while the latter was formed by high-ranking Nationalist figures who had broken with the Nationalists.

Given this situation, the Nationalist government sought to engage with the Hong Kong government over the extradition of key opposition figures such as Li Jishen 李濟深 (Li Cai Sum) and Xu Jizhuang 徐繼莊 (Hsu Chi Chuang). Li Jishen was a prominent anti-Chiang General in the Guomindang and a co-founder of the GRC.[104] He was accused by the Nationalists of using the colony as a recruiting base for an armed uprising against the Chinese government. Chiang personally raised the matter with Stevenson but without an effective outcome.[105] For his part, Xu Jizhuang was a former staff-member of the Central Bank of China. He had fled to Hong Kong after having been accused of committing gross embezzlement.[106] But a Hong Kong court deemed that the evidence that the Guomindang had submitted was short of what was required and thus refused to issue an extradition order.[107] Nevertheless such issues added further distractions to the quest for restoring Hong Kong to China.

---

**103** Minute by China Department on the Communist threat to Hong Kong (F 15770/154/10), October 19, 1948, Ashton, Bennett and Hamilton, eds., *DBPO, S.1, V. 8: Britain and China 1945– 1950*, 163–164.

**104** Other major co-founders were Madam Sun Yat-sen (Song Qingling 宋慶齡) and He Xiangning 何香凝, Liao Zhongkai's 廖仲愷 widow, both on the left wing of the Nationalist Party.

**105** TNA, FO 371/63388, telegram from Sir R. Stevenson to Hong Kong, May 8, 1947; TNA, FO 371/ 63388, telegram from Hong Kong to the Colonial Office, May 21, 1947.

**106** TNA, FO 371/69582B, aide-memoire by the Chinese Embassy in London, August 9, 1948.

**107** TNA, FO 371/69582B, telegram from A.L. Scott (FO) to J.B. Sidebotham (CO), August 10, 1948.

## 5 Further Conflict between China and Britain: the Kowloon Walled City, 1948–1949

The signature of the two aforementioned agreements with the Hong Kong government indicated the Guomindang's acknowledgement of British treaty rights in the colony but did not dispel local anxieties about the colony's future. A clamour for the reversion of Hong Kong and Macao occurred periodically in the Chinese press. The Portuguese authorities in Macao wished to establish a "united front" with the British, but the latter refused to entertain this.[108] To the Foreign Office's eyes, "Hong Kong is a much sounder proposition than Macao."[109] Macao had never been ceded to the Portuguese and the Sino-Portuguese Treaty of 1888 merely confirmed "the perpetual occupation and government of Macao and its dependencies by Portugal."[110] The Portuguese thus chose to go it alone and stress their determination to retain Macao.[111] The Foreign Office also kept on declining the Colonial Office's request for taking a public stance.[112] Ernest Bevin explained to Arthur Creech Jones, the Colonial Secretary, that

> We know that it is the aim of the Chinese government to regain Hong Kong eventually, but in present circumstances they are prepared to let these ambitions lie dormant. If we volunteer a statement on the future of the Colony, there is a considerable risk that the Chinese may be stung to action.[113]

Yet despite the Foreign Office's cautious position, the British authorities in Hong Kong inflamed the situation, having resolved to remove Chinese influence in the Kowloon Walled City. The city had been more or less destroyed during the war and the city wall had been demolished in 1942 by the Japanese to obtain materials for the extension of the Kai Tak aerodrome. Only two buildings remained intact: a dilapidated former school and a home for aged women.[114] In 1946–1947, the city had become a home for Chinese squatters who erected 54 wooden huts. Following Chinese attempts to restore their jurisdiction over the city, the Nationalist presence in the city had gradually increased and had now become

---

108 TNA, FO 371/63390, telegram from A.L. Scott (FO) to Sir Orme Sargent (FO), August 12, 1947.
109 TNA, FO 371/63389, telegram from G.V. Kitson (FO) to Sir Ralph Stevenson, July 17, 1947.
110 TNA, FO 371/63390, telegram from G.V. Kitson (FO) to L.H. Lamb (Nanjing), July 29, 1947.
111 TNA, FO 371/63390, telegram from Sir Ralph Stevenson to G.V. Kitson (FO), July 3, 1947; TNA, FO 371/63392, letter from N.L. Mayle (CO) to A.L. Scot (FO), November 20, 1947.
112 TNA, FO 371/63392, letter from A.L. Scott (FO) to N.L. Mayle (CO), December 4, 1947.
113 TNA, FO 371/69582A, letter from Ernest Bevin to Arthur Creech Jones, June 9, 1948.
114 TNA, CO 537/1657, telegram from Governor of Hong Kong to Colonial Office, September 24, 1946.

a source of instability in British eyes. On the grounds of taking precautions against the threat of fire and disease, the Hong Kong government thus demanded that all squatters leave the city by December 11, 1947. Then, on January 5, the government dismantled all the huts. Lin Xiazi, the magistrate of Bao'an County, immediately intervened and encouraged the Chinese to remain in the city. Some huts were subsequently re-erected on the former sites. The Hong Kong government decided to demolish these newly built huts on January 12. This caused a severe clash between the police and the Chinese crowd. One Chinese was killed, one suffered a serious abdominal wound caused by a bullet and five others suffered minor injuries.[115]

The Nationalists strongly condemned this incident, but confined the matter to the matter of the Kowloon Walled City rather the wider future of Hong Kong.[116] Yet Chinese public fury spiralled out of the Guomindang's control.[117] The British Consulate in Guangzhou and the commercial premises of Butterfield and Swire were attacked and completely burned out, as well as part of the buildings of the Chartered Bank and the Jardine Matheson company.[118] This violence was later termed by historians "the 1948 Shamian Incident" (*Shamian shijian* 沙面事件) as these buildings were located on Shamian Island. The British asked the Chinese government to compensate the British for the damage, but the latter insisted that the sovereignty of the Kowloon Walled City should be settled first.[119] Wang Shijie believed that Chinese claims over the city were weak in legal terms and its recovery should not be the Nationalists' prime diplomatic target. In his mind, the targets should be the entirety of Kowloon and Hong Kong.[120] Nonetheless, he had to balance this against the public pressure to act. The Kowloon Walled City thus became a focus of Sino-British negotiations in 1948.

---

**115** TNA, CO 537/3705, "Note on eviction from Kowloon Walled City" by Colonial Office, January 19, 1948.

**116** TNA, CO 537/3705, telegram from Sir R. Stevenson to Foreign Office, January 10, 1948.

**117** According to Zhang Junyi's research based upon Song Ziwen's papers, this Chinese protest was badly handled by the Guangzhou branch of the Guomindang. To make capital of this event, and also to prevent the CCP's involvement, members in the Guangzhou branch presided over the demonstration but failed to rein the protestors back from violence due to poor cooperation between various factions inside the Guomindang. See Junyi Zhang, "1948 nian Guangzhou Shamianshijian shimo: yi Song Ziwen dang'an wei zhongxin [The History of the 1948 Shamian Incident in Guangzhou: A Study Based on the T.V. Soong Papers]," *Social Sciences in China*, no. 6 (2008): 185.

**118** TNA, CO 537/3705, telegram from Hong Kong to Colonial Office, January 17, 1948.

**119** TNA, CO 537/3707, telegram from Sir R. Stevenson to Foreign Office, July 19, 1948.

**120** January 8, 1949, see Lin, ed., *The Diary of Dr. Wang Shih-chieh*, 884–885.

Several formulae were devised to try and solve this Sino-British dispute. There was, for example, a proposal to establish a Chinese Maritime Customs post in the city, but this was opposed by the Hong Kong authorities as a surrender to the Nationalists.[121] There was also dialogue over the creation of a local office of the Chinese Special Commissioner for Foreign Affairs relating to Guangdong and Guangxi in the city, yet this too was rejected by the Hong Kong administration.[122] Further consultations took place over the location of a Chinese consulate compound in the city, but this met with a similar fate.[123]

Another innovative idea was the designation of the city as a Garden of Remembrance. The Colonial Office proposed that this should be dedicated to all those, irrespective of nationality, who had given their lives in the war against the Japanese.[124] The Chinese presented no initial objection to this plan but demanded that the care and maintenance of the park should be entrusted to the Chinese government. The British, in turn, suggested the administration of the garden should be jointly undertaken by two trustees, one Chinese and one British. This was agreed by the Chinese on the condition that Britain make a public announcement to the effect that the whole area was under Chinese jurisdiction. However, the British government was not prepared to do this. In any event, the Hong Kong government had always deemed the co-administration of the garden unworkable.[125]

Meanwhile, as the Chinese civil war approached Shanghai in late 1948, the United Nations Economic Commission for Asia and the Far East (UNECAFE)[126] Secretariat was preparing to leave that city. The Hong Kong government took the opportunity to suggest that the whole of the Kowloon Walled City area might be handed over jointly by the British and the Chinese to the United Nations for the use of the permanent secretariat of the UNECAFE or some other

---

**121** TNA, CO 537/3705, letter from M. E. Dening (FO) to G.F. Seel (CO), January 15, 1948; TNA, CO 537/3707, table of "Kowloon City Summary of Negotiations", assumedly in late 1948.

**122** TNA, CO 537/3705, telegram from Foreign Office to Nanjing, January 27, 1948; minute to Mr Steel by N.L. Mayle, February 5, 1948; and telegram from Hong Kong to Colonial Office, February 3, 1948.

**123** TNA, CO 537/3705, minute to Mr Steel by N.L. Mayle, January 30, 1948; TNA, CO 537/3705, telegram from Hong Kong to Colonial Office, February 3, 1948; TNA, CO 537/3707, table of "Kowloon City Summary of Negotiations", assumedly in late 1948.

**124** TNA, CO 537/3705, minute to Mr Steel by N.L. Mayle, February 6, 1948.

**125** TNA, CO 537/3707, table of "Kowloon City Summary of Negotiations", assumedly in late 1948.

**126** The Commission was renamed in 1974 the Economic and Social Commission for Asia and the Pacific (ESCAP).

UN organisation.[127] The Foreign Office objected, however, since in its view, the city could not provide immediate accommodation for the Secretariat on its evacuation from Shanghai.[128] As things transpired, the UNECAFE eventually relocated to Bangkok.

As a further option, the Hong Kong government advocated an approach to the International Court of Justice and that in the interim the status quo of the city should be maintained.[129] But legal experts in London were hesitant about this:

> in general, we do not feel that the British case really is in any sense impregnable. Apart from the defence of military requirement we think that the British case is weak, and we feel a little uncertainty ... whether ... it could be affirmatively established that there was some actual military situation at present subsisting or likely to develop in the foreseeable future which made it impossible to concede the Chinese claim to resume jurisdiction.[130]

The Hong Kong government continued to press this case. It believed that such a step would impress Hong Kong residents that the British government was indeed serious about retaining the colony; otherwise, the Hong Kong Chinese might be more reluctant to back the British side.[131] Yet the authorities in London took the advice of the legal experts and came to the conclusion that not only was it ill-advised to take the case to the International Court but that Hong Kong government "must also avoid getting [itself] into a position where the Chinese might take the case to the Court"; "it is highly desirable that some compromise solution should be arrived at [between Britain and China]."[132] The Foreign Office recommended that the Colonial Office accept the view of Ralph Stevenson, namely that "the issue [of the Kowloon Walled City] might be left dormant at this juncture."[133]

In February 1949, Guo Dehua submitted a final Nationalist proposal that "authorities in Hong Kong should admit Chinese sovereignty over this small area in return for which the Chinese government should immediately ask the Government of Hong Kong to assume administrative responsibility."[134] The British government, however, was now in no mood to listen to a regime which was shortly

---

127 TNA, CO 537/3707, telegram from Hong Kong to Colonial Office, November 24, 1948.
128 TNA, CO 537/3707, letter from J.F. Turner (FO) to Miss J. Burnham (CO), November 29, 1948.
129 TNA, CO 537/3706, telegram from Hong Kong to Colonial Office, March 14, 1948.
130 TNA, CO 537/3707, report by Law Officers' Department, June 24, 1948.
131 TNA, CO 537/3707, telegram from Hong Kong to Colonial Office, July 27, 1948.
132 TNA, CO 537/3707, telegram from Colonial Office to Hong Kong, August 11, 1948.
133 TNA, CO 537/3707, letter from P.W. Scarlett (FO) to J.B. Sidebotham (CO), December 16, 1948.
134 TNA, CO 537/4807, letter from G.F. Tyrrell (Guangzhou) to C.B.B Heathcote-Smith (Hong Kong), February 26, 1949.

going to collapse.[135] The British and the Chinese were thus unable to find a compromise over the sovereignty of the Kowloon Walled City.

## 6 Britain's Hong Kong Policy in the Face of the Transition of Power in China, 1948–1949

At the end of 1948, British policy was not much changed since the position statements laid out by the Foreign Office and the Colonial Office in 1946–1947. In October, J.B. Sidebotham, an official in the Colonial Office, recommended

a. that we shall not agree to hand back the New Territories to China until the end of the lease; but
b. that, if for any reason it was decided to do so, we should, as part of the deal with a friendly China, be able to make conditions which would ensure access to the water supplies and, possibly, even some contribution towards the capital cost;
c. that, in the case of all development in the New Territories, we should be careful to plan it on the basis that the capital cost will be fully amortized upon the end of the leases; and
d. that continued development in the New Territories is desirable as an indication to the Chinese that we intend to remain there until the end of the lease.[136]

Yet Sidebotham also acknowledged that these approaches would not be feasible unless a friendly power existed in China.[137] When it came to the question of whether it was practical for the British to hold on to Hong Kong once the New Territories were again in Chinese hands, he felt that this depended on two conditions:

a. whether our relations with China are then on a sufficiently friendly basis for them to be willing that we should remain there for the mutual benefit of China and ourselves; or
b. whether, even short of active aggression, [the Chinese] could make the continued existence of Hong Kong as a British territory impossible.[138]

Sidebotham's note implied that the fate of Hong Kong was to a large extent dependent on the Chinese rather the British.

---

135 TNA, CO 537/4807, letter from Leo H. Lamb (Nanjing) to C.B.B Heathcote-Smith (Hong Kong), March 24, 1949.
136 TNA, CO 537/3712, note by J.B. Sidebotham, October 11, 1948.
137 Ibid.
138 Ibid.

Meanwhile, debates between British colonial and diplomatic officials continued over a reassuring public statement about Britain's determination to retain Hong Kong. The Governor of Hong Kong stressed that "we should stay as long as we can and, if that is our intention, it would steady the minds of people in the Colony if we could say so."[139] He proposed that an unambiguous statement could be made in the House of Commons through arranging for an appropriate question to be asked:

> Will the Secretary of State [for Colonial Affairs] say whether it is HM Government's intention that Hong Kong will remain British?
> The answer suggested is 'Yes'.[140]

Stevenson thought that it would be better to ask a more general question about Southeast Asia. In his opinion, "the fewer statements on Hong Kong the better."[141] It was, however, eventually decided that no such a question would be put in the House of Commons.[142]

Back in Hong Kong, with the Nationalist government under pressure in Guangdong province from the CCP troops, the government of the colony pulled down shacks in the Kowloon Walled City. It was reported that 2,000 shacks, occupied by 6,000 Chinese squatters, were razed on "public health" grounds on July 3, 1949 under police supervision.[143] Thereafter, regular uniformed Hong Kong patrols were introduced into the city.[144] Peter W. Scarlett, the head of Far Eastern Department (1947–1950), asserted that if formal protests were made by the Nationalists "we may with luck be able to stall until they finally disappear."[145] Yet anxieties in London about Chinese reactions now also derived from the Communist side. The Colonial Office instructed the Hong Kong government to desist from evicting Chinese squatters for fear of agitating Chinese Communist leaders.[146] In the Kowloon Walled City allegiance soon shifted to the Chinese Communists. On October 16, the local residents association hoisted a Communist flag and sent a congratulatory message to Mao Zedong, the Chair-

---

139 TNA, FO 371/75839, minute by Sir W. Strang (FO), March 1, 1949.
140 TNA, FO 371/75839, telegram from Hong Kong to Colonial Office, January 26, 1949.
141 TNA, FO 371/75839, telegram from Sir R. Stevenson to Foreign Office, February 18, 1949.
142 The question did not appear in the debates in the House of Commons in 1949. See https://hansard.parliament.uk/Commons.
143 TNA, CO 537/4807, newspaper article clipped from *The New York Times* of July 4, 1949.
144 TNA, CO 537/4807, extract from the Hong Kong government's telegram to Colonial Office, September 10, 1949.
145 TNA, CO 537/4807, letter from P.W. Scarlett (FO) to J.B. Sidebotham (CO), September 2, 1949.
146 TNA, CO 537/4807, telegram from Colonial Office to Hong Kong, August 27, 1949.

man of the PRC. The head of the association said they would look to the new Bao'an county government for instructions.[147]

In mid-1949, the British government was still unable to ascertain the CCP's Hong Kong policy, although Mao had made his views about the colony known via a British journalist at the end of 1946, maintaining that the CCP did not seek an early return of Hong Kong. This was reiterated by Qiao Guanhua 喬冠華, the CCP's *de facto* representative in Hong Kong, in late 1948 when he stressed that "it was not the Communist Party's policy to take the British colony by force when they come into power in China."[148] Yet an incident puzzled the British. On April 20, 1949 Communist troops shelled a British frigate, HMS *Amethyst, en route* from Shanghai to Nanjing to take supplies to the British Embassy in Nanjing. The captain was injured and soon died of his wounds. It did not seem like an accident since the British rescue ships were subsequently attacked and suffered serious damage.[149] This British naval setback on the Yangtse was a blow to British prestige in Hong Kong. It was strongly suspected in Chinese communities there that the British did not have enough strength to defend Hong Kong if the Communists pressed for rendition or sought to capture the colony.[150] To repair lost confidence, British military reinforcements were sent to beef up the garrison in Hong Kong, increasing the garrison from a few thousand to 30,000 and adding tanks, fighter aircraft and a naval unit.[151]

Concurrently, as the Cold War began to spread across East Asia, the attitude of the USA towards Hong Kong quietly changed. Dean G. Acheson, the US Secretary of State, found no reason to challenge the British viewpoint that it was in their common interests for the United States and Britain to stand firm over Hong Kong since the city was "a useful window" and enabled "us to exercise influence on the Chinese situation" as well as gather intelligence about China.[152] Furthermore, Hong Kong would facilitate business dealings with China as it was an efficiently-run port and a hub of international trade in East Asia. Nevertheless, the Unites States in the end decided against giving direct assistance to Hong Kong's defences on the grounds that to defend Hong Kong would require

---

147 TNA, CO 537/4807, extract from the Hong Kong government's telegram to Colonial Office, October 21, 1949.
148 Tsang, *A Modern History of Hong Kong*, 152–153.
149 Loh, *Underground Front*, 77.
150 TNA, FO 371/75839, telegram from Hong Kong to Colonial Office, April 30, 1949.
151 Loh, *Underground Front*, 78.
152 TNA, FO 371/75839, telegram from Sir O. Franks (Washington) to Foreign Office, April 2, 1949.

the establishment of a military position well inland and this would incite a global war.[153]

In these circumstances, the Colonial Office hoped that the future of Hong Kong was still a matter which could be negotiated with a new government in China and that this could be facilitated by an announcement of a desire to create amicable relations with whatever government emerged there.[154] In the interim, the British government's position in September 1949 was that:

> We should be prepared to discuss [the] future of Hong Kong with a friendly and stable government of a unified China, [but] conditions under which such discussions could be undertaken do not exist at present and are unlikely to exist in foreseeable future.[155]
>
> The lease of the New Territories expires in 1997. It does not seem likely that any Chinese Government will be prepared to renew [the] lease. Without these territories Hong Kong would be untenable and it is therefore probable that before 1997 the United Kingdom Government of the day will have to consider [the] status of Hong Kong. It is not possible, however, some two generations in advance to lay down principles which should govern any arrangement which it may be possible to reach with China at that time. A decision at [the] present time can therefore only be taken on an interim policy.[156]

Hence, the chief task for the British was to postpone the "evil day" for as long as they could.[157] Yet ensuing developments soon began to dissipate their apprehension. It transpired that the CCP had not fundamentally altered its Hong Kong policy. Communist troops approaching Hong Kong's border were specifically instructed to show restraint and avoid conflict.[158] After the outbreak of the Korean War, Hong Kong's value as a channel to avoid the UN and US embargoes became prominent for the PRC. Zhou Enlai, the Chinese Premier, stressed the colony's usefulness as "a valuable instrument ... to divide the British from the Americans in their East Asian policies".[159] As a consequence, the recovery of Hong Kong was shelved by the Chinese Communist government until the British

---

153 Tsang, *Hong Kong: An Appointment with China*, 76.

154 TNA, FO 371/75839, letter from J.J. Paskin (CO) to Mr. Scarlett (FO), June 24, 1949.

155 TNA, FO 371/75839, outward telegram (No. 325) from Commonwealth Relations Office, September 7, 1949.

156 TNA, FO 371/75839, outward telegram (No. 326) from Commonwealth Relations Office, September 7, 1949.

157 Tsang, *Hong Kong: An Appointment with China*, 77.

158 Sheng Zeng, *Zeng Sheng huiyilu [The Memoir of Zeng Sheng]* (Beijing: Jiefangjun Chubanshe, 1991), 570–571.

159 Jiatun Xu, *Xu Jiatun huiyilu [The Memoir of Xu Jiatun]*, 2 vols., vol. 2 (Hong Kong: Lianjing Chunban Shiye Jituan, 1993), 473–474. Quoted from Tsang, *A Modern History of Hong Kong*, 154.

revived it in the 1980s. In 1997 the whole area of Hong Kong was finally returned to China.

## Conclusion

The Chinese Nationalist Party's Hong Kong policy during 1946–1949 was entangled with its domestic concerns. On the one hand, it was the Nationalist government's policy to avoid any formal discussions about Hong Kong's future before internal hostilities with the Communists ceased. On the other, the Nationalist authorities had to adopt an outwardly tough stance in dealing with any troubles derived from Hong Kong under pressure from the Chinese public, as well as some Nationalist officials. As a result, the Guomindang's manoeuvres over Hong Kong looked ambivalent. For instance, it took pains to quieten down tensions arising from Hong Kong while creating new disputes with the British by, for example, acquiescing in local officials' pursuit of the resumption of Chinese jurisdiction over the Kowloon Walled City. By 1947–1948, domestically economic and financial concerns outweighed the Nationalist quest for resuming sovereignty over Hong Kong. Against this background, the government decided to reaffirm British treaty rights in Hong Kong in exchange for the latter's cooperation in anti-smuggling and anti-speculation efforts. However, this did not end Sino-British conflicts over Hong Kong. Chinese public outrage, driven by the Hong Kong government's inflammatory measures in evicting Chinese squatters from the Kowloon Walled City, resulted in an outbreak of violence in January 1948. Differences over the issue of the Kowloon Walled City then frustrated the Guomindang's relations with the British in the final year of its rule in mainland China.

On the British side, the transition of power from the Conservative Party to the Labour Party in Britain brought about very few changes in Britain's Hong Kong policy. Its policy in 1946–1949 still swayed between the opinions of the Foreign Office and the Colonial Office. The former was fully aware of the Nationalists' desire to regain Hong Kong but also of the domestic predicament that the Chinese faced in achieving this goal. Therefore, it recommended avoiding any actions that might antagonise the Chinese and provoke them into formal discussions over Hong Kong's future. By contrast, the Colonial Office, siding with the Hong Kong government, was more concerned about fragile British prestige in Hong Kong. From its perspective, the fervent nationalism of the Chinese population after the war, and the Nationalist manoeuvres over local issues in Hong Kong, such as the airfield construction in the New Territories and jurisdiction over the Kowloon Walled City, hindered British rule. Although the two departments agreed on the inevitability of British concessions over Hong Kong, their disagreements continued over the details of such a compromise and, particularly, whether to publicly assert British intentions to remain in Hong Kong. In the end, the British simply sought to postpone any decision to a later date. But the fate of Hong Kong was fraught with uncertainties.

It was apparent that Hong Kong, after the abolition of the treaty-port system and extraterritoriality in 1943, and in view of the fading importance of Tibet in British eyes owing to the latter's retreat from India in 1947, gradually became the focus of British post-war concerns over China. Although the Nationalist government raised no formal discussion in 1946–1949, the issue of the colony dominated the Sino-British relationship during this period. The Sino-British tensions over Hong Kong not only strengthened Chinese anti-British and anti-imperialist sentiments, but also convinced the British government that a harmonious Sino-British relationship was vital whether to retain Hong Kong or not. In this sense, the fate of Hong Kong had already fallen into Chinese hands. This, to some degree, also affected the later British attitude towards the Chinese Communist government and contributed to its formal recognition of the latter in December 1949, as the first country to do so in the Western Bloc.

# Chapter V
# Tibet and the Wartime Sino-British Relationship, 1942–1944

This chapter examines Sino-British conflicts over the opening of a trans-Tibet supply route in 1941–1943 and the resultant British re-examination of its Tibetan policy. Several historical works have previously studied these matters, but their research either heavily relied on single-language documental sources or put emphasis on one party's thoughts and practice, either Chinese or British.[1] Consequently, they are short at giving equal treatment to both sides in the Sino-British-Tibet dynamics and produced some misinterpretations, for example in regard to the Chinese motive behind militarily warning the Lhasa government in early 1943 and Nationalist attitudes towards opening a trans-Tibet animal pack route by way of a China-Tibet-Britain tripartite accord. This chapter endeavours to overcome these shortcomings. As the chapter will show, the impact of the Second World War upon the Sino-Tibet-Britain relationship was significant. The need for a new supply route to China not only rekindled debates over Tibet's sovereignty but also this matter to the attention of United States. It further spurred the British-Indian government into a *de facto* control of the disputed Indo-Tibetan borderlands for fear that the Chinese would claim these territories once Tibet fell into the Chinese orbit after the war. As a result, it was not only the Sino-British relationship that suffered, but also its Anglo-Tibetan and Indo-Tibetan counterparts. All these changes were sparked by the Allied nations' support of China's war effort from India via Tibet.

---

**Acknowledgement:** This chapter is derived in part from an article of mine published in The International History Review, August 17, 2021, copyright Taylor & Francis, available online: http://www.tandfonline.com/10.1080/07075332.2021.1921828

1 Such works mainly depending on English-language documents are *Lamb, Tibet, China & India, 1914–1950; ibid., McMahon Line (I); ibid., McMahon Line (II); ibid., The China-India Border: the Origins of the Disputed Boundaries* (London, New York and Toronto: Oxford University Press, 1964); *Goldstein, Demise of the Lamaist State; Smith Jr., Tibetan Nation; Singh, Himalayan Triangle; Guyot-Réchard, Shadow States.* Those works that use Chinese-language sources always consult English-language documents as a supplement but their focuses are generally laid on the Chinese side, such as *Chen, Kangzhanqianhou zhi Zhong-Ying Xizang jiaoshe; Zhu, Xizang zhuanshi; Feng, Zhong-Ying Xizang jiaoshe yu Chuanzang bianqing; Tuttle, Tibetan Buddhists in the Making of Modern China;* and *Lin, Tibet and Nationalist China's Frontier.* To be noted, Goldstein's and Zhu's works also consult Tibetan-language sources.

https://doi.org/10.1515/9783110706659-012

# 1 China, Britain and Tibet before 1942

The triangular Anglo-Chinese-Tibetan relationship can be traced back to the time when the British East India Company took control of Bengal in 1765 and subsequently expanded its influence towards the Himalayan areas. At that time, the territories along the Himalayas, such as the tribes and kingdoms of Ladakh, Lahaul, Spiti, Garwhal, Kumaon, Nepal, Sikkim, Bhutan and Assam, were in the sphere of Tibetan Buddhism and culture.[2] In 1861 and 1865, Britain waged two wars in Sikkim and Bhutan, two vassals of Tibet, and placed them under British influence. This, however, brought the British into potential conflict with Tibet and China.

## 1.1 Tibet's Position in the Qing Empire, 1648–1903

The Qing's control over Tibet can be traced back to 1648 when the Shunzhi 順治 Emperor received the Fifth Dalai Lama[3] at Beijing. This has been interpreted as Tibet's formal submission to the Qing Empire.[4] Yet effective control over Tibet was not established until the Kangxi 康熙 Emperor expelled intervention by the Dzungar (Zhunga'er 準噶爾) Mongols from Tibet in 1720. After that, Tibet was incorporated into the Qing Empire. Its relations with the Qing court had both religious and political aspects.

There was a close religious link between Qing imperial elites and Tibetan Buddhist communities, generally termed as the patron-priest relationship (*tanyue guanxi* 檀越關係 or *gongshi guanxi* 供施關係). In contrast to the Han Chinese who were multireligious, with Daoism and Buddhism incorporated into the cosmos of popular religion, many Manchus and Mongols were followers of Tibetan Buddhism. After establishing their authority in China, the Manchu rulers became the primary patrons of monasteries and lamas in Tibet in trade for the latter's blessing and spiritual protection. Tibetan Buddhism was also exploited by the Manchus to cement their rule over Mongols and strengthened the Qing government's control over Mongolia.[5] Tibetan monasteries and lamas thus thrived in Beijing, Rehe 熱河, the Qing's summer capital, and Wutaishan 五臺

---

2 Lamb, *British India and Tibet, 1766–1910*, 3.
3 The Dalai Lama was the foremost religious figure in the Gelukpa (Gelupai 格魯派) school of Tibetan Buddhism and the Panchen or Tashi Lama was the second.
4 Smith Jr., *Tibetan Nation*, 108–113.
5 Tuttle, *Tibetan Buddhists in the Making of Modern China*, 31–33.

山 acting as Lhasa's intermediaries at the Qing court.[6] Nevertheless, it is too narrow to interpret the religious link merely from economic and political perspectives. There were indeed some Buddhist connections. Tibetan Buddhists, for example, worshipped Qing emperors as manifestations of Manjushri (Wenshupusa 文殊菩薩), a bodhisattva of the same rank with, or slightly higher than, that of the Dalai Lama, Avalokiteshvara (Guanshiyinpusa 觀世音菩薩), and that of the Panchen Lama, Amitābha (Amituofo 阿彌陀佛), according to Buddhist scriptures.[7] The worship contributes to the prosperity of Tibetan Buddhism at Wutaishan as the mountain is the Daochang 道場 (Bodhimaṇḍa) of Manjushri, a seat where Manjushri's enlightenment is present according to Chinese and Tibetan Buddhism.

Politically, Tibet was subservient to Beijing. Beijing's power was mainly projected over central Tibet (Weizang 衛藏 or Ü-Tsang) through two prominent figures of the Gelukpa school: the Dalai Lama and the Panchen Lama. Like its Yuan 元 and Ming 明 predecessors, the Qing government was expert at manipulating competition between different Tibetan Buddhist schools. Under the Qing court's endorsement, the Gelukpa school consolidated its dominance over central Tibet and the Dalai Lama, the most important incarnate lama of the school, rose to be the supreme figure of Tibetan Buddhism and the nominal ruler of central Tibet.[8] To balance the Dalai Lama's absolute power, some territorial and political divisions were introduced. The areas under the Dalai Lama's direct domain were confined to Ü or "Hither Tibet" (Qianzang 前藏) while the Panchen Lama, the incarnate lama subordinated only to the Dalai Lama in the hierarchy of the Gelukpa school, assumed temporal authority over Tsang or "Further Tibet" (Houzang 後藏), including Tibet's second largest city, Shigatse (Rikaze 日喀則).[9] In the case of a Dalai Lama's minority, a temporary office of the regent (shezheng 攝政) would manage the government. Below the Dalai Lama was the Kashag (gaxia 噶廈), a council of ministers with executive functions. The council consist-

---

**6** Ibid., 19 – 25.
**7** Many Yuan and Ming emperors are also worshipped by Tibetan Buddhists as manifestations of Manjushri. This tradition of regarding rulers of China proper as Chakravartins (Zhuanlunwang 轉輪王) or incarnations of Manjushri also extends to Mao Zedong, the Chairman of the People's Republic of China. See Haiyan An, "Zuowei 'Zhuanlunwang' he 'Wenshupusa' de Qingdi: jianlun Qianlongdi yu Zangchuanfojiao de guanxi [Emperor as Chakravartin and the Incarnation of Mañjuśrī: the Relationship between the Qianlong Emperor and Tibetan Buddhism]," *Qingshi Yanjiu [The Qing History Journal]*, no. 2 (2020): 106 – 109.
**8** Tuttle, *Tibetan Buddhists in the Making of Modern China*, 18 – 19.
**9** Smith Jr., *Tibetan Nation*, 130. Smith argues that the temporal authority of the Panchen Lama extended to Ngari (Ali 阿里), whereas in fact the place was directly ruled by the Kashag under the Qing court's supervision.

ed of four ministers or Kalons (galun 噶倫), one of these being a lama and three secular officials. At the local level, power remained in the hands of monasteries and some aristocrats. The Tibetan military was distributed across various regions, leaving the most important strategic locations, such as Lhasa, under imperial garrisons.[10]

The Qing court's authority over central Tibet culminated in the late eighteenth century but declined gradually afterwards. From 1728, two high-ranking imperial officials (Manchu: amban, Chinese: dachen 大臣) appointed by the central government resided in Lhasa and in 1792 both the Dalai Lama and the Kashag were officially placed under their broad supervision.[11] Also, as part of the 1792 reforms, Tibetan frontier defences and foreign affairs fell within the ambans' jurisdiction, as did the command of the Tibetan troops, the right to review all judicial decisions and the supervision of the Tibetan currency. Most significantly, the Qing's authority entered the religious realm. The reincarnation of the Dalai Lama, the Panchen Lama, and other high-ranking incarnate lamas, was to be supervised and presided over by the ambans. A lottery system (*jinpingzhiqian* 金瓶挚签) was introduced to decide the high-ranking lamas and prevent these positions being monopolised by powerful Tibetan families. At this lottery ceremony, the imperial amban picked a name from several lots placed in a Golden Urn (*jinbenbaping* 金賁巴瓶).[12] Although the bulk of these arrangements was maintained until the collapse of the Qing Dynasty, Beijing's rule over Tibet gradually weakened throughout the second half of the nineteenth century when it frequently suffered from external encroachment and internal disorder. The Thirteenth Dalai Lama, for example, was not selected by lottery in 1878.[13]

## 1.2 The British Invasion of Lhasa and the Qing Court's Modern Nation State Building in Tibet, 1904–1911

Before 1904, the British government had never challenged China's authority in Tibet. Trade privileges regarding Tibet and the settlement over the Tibet-Sikkim boundary were obtained from treaties with the Qing government without Lhasa's

---

**10** Luciano Petech, *China and Tibet in the Early 18th Century: History of the Establishment of Chinese Protectorate in Tibet* (Leiden: E.J. Brill, 1950), 223–235.

**11** Ibid., 236–237; Smith Jr., *Tibetan Nation*, 130–131.

**12** Zhou Hongwei, *Yingguo, Eguo yu Zhongguo Xizang [Britain, Russia and Chinese Tibet]* (Beijing: Zhongguo Zangxue Chubanshe, 2000), 13–17.

**13** Goldstein, *Demise of the Lamaist State*, 44.

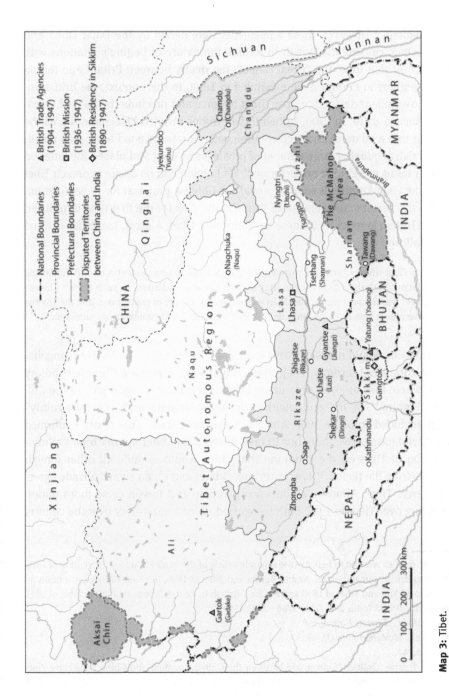

**Map 3:** Tibet.
© Zhaodong Wang & © Peter Palm, Berlin.

involvement.[14] However, these agreements were defied by the Dalai Lama and the Kashag, causing the British to rethink the nature of Beijing's relations with Lhasa.[15] At that time, the "Great Game", the rivalry between Britain and Russia for hegemony in Central Asia, expanded to Tibet. In this context, the British-Indian government decided to open Tibet by force and preclude the alleged Russian influence there. In 1904, British-Indian troops invaded Lhasa and forced the Kashag into a bilateral Lhasa convention without reference to China. As a result, the British acquired a large number of privileges in Tibet and placed the latter on similar terms to a British protectorate.[16] Yet London had no desire to detach Tibet from the Qing Empire. It soon accepted the Chinese proposal for re-negotiation. In the discussions, the concept of "suzerainty" was for the first time introduced by the British to redefine Chinese authority over Tibet. Alastair Lamb has defined it as follows:

> In a Tibet where China was suzerain the British could to some extent establish diplomatic relations with the Tibetan authorities without Chinese mediation. In such a Tibet Chinese power was not very much more than a ceremonial reminder of past glories, and the Tibetan government was in its dealings with British India able to demonstrate its autonomy.[17]

But the Qing would not accept this term since, in their eyes, Tibet, like Mongolia, was a conquered territory. They stressed that Tibet was not a Chinese tributary state where they enjoyed only limited rights: it was an integral part of the Chinese Empire.[18] In the end, neither the term "sovereignty" nor "suzerainty" were included into the 1906 Anglo-Chinese agreement. This treaty confirmed the 1904 Lhasa convention, but Britain agreed not to interfere in China's administration of Tibet on the condition that China would exclude all other foreign powers from the region.[19] Two years later Britain and China signed a trade agreement over Tibet. Almost simultaneously, Britain and Russia came to an understanding over Tibet in 1907. Both recognised China's suzerainty over the country

---

14 The treaties referred to here are the separate article of the Yantai (Chefoo) convention of 1876, the Sino-British convention relating to Burma and Tibet of 1886, the Sino-British convention relating to Sikkim and Tibet of 1890 and the Sino-British trade regulations relating to Tibet of 1893.
15 Lamb, *British India and Tibet, 1766–1910*, 114–192.
16 Smith Jr., *Tibetan Nation*, 159.
17 Lamb, *McMahon Line (I)*, 45.
18 Ibid., 44.
19 UK Treaties Online, "Convention between the United Kingdom and China respecting Tibet (To which is annexed the Convention between the United Kingdom and Tibet, signed at Lhasa, September 7, 1904)," 1906, http://foto.archivalware.co.uk/data/Library2/pdf/1906-TS0009.pdf (accessed April 14, 2019).

and agreed not to interfere in Tibet's internal affairs or to negotiate with Tibet except through the Chinese government.[20]

Although the Qing court's sovereignty over Tibet was reaffirmed, Beijing was alarmed by the potential separation of Tibet from the Qing Empire. To safeguard the Qing's territorial integrity and also invoke the western concept of sovereignty, the Qing government decided to establish its absolute authority over Tibet.[21] In 1906, Zhang Yintang 張蔭棠 (Chang Yin-tang) was sent to reorganise the Tibetan government and abolish the Dalai Lama's temporal power.[22] In tandem with this, Zhao Erfeng 趙爾豐 (Chao Erh-feng) was instructed to impose the replacement of local hereditary headmen with central government-appointed officials (*gaitugui-liu* 改土歸流) on the Kham areas (Kangqu 康區), comprised of western Sichuan and eastern Tibet, with the purpose of detaching a new province from Tibet, named Xikang 西康, which was placed under the Qing court's direct control. To back up these plans, 2,000 Chinese troops advanced into Lhasa in 1910. This reform, however, exacerbated Beijing's relations with the Thirteenth Dalai Lama and drove the latter to flee to British India. In his ensuing two-year exile in India, this once anti-British figure gradually became more pro-British.[23] Nevertheless, the British had to admit that the Chinese had regained "full control" over Tibet between 1907 and 1911.[24]

## 1.3 The De Facto Independence of Tibet and the Simla Conference, 1912–1914

The Qing government's efforts to impose greater control over Tibet soon failed due to the 1911 Revolution and the subsequent collapse of the Qing Dynasty.[25] After resuming power in Lhasa, the Thirteenth Dalai Lama started his pursuit of an independent and central-administrative Tibetan nation-state. The historical Sino-Tibetan relationship was reinterpreted, that China was not a suzerain to a subordinate Tibet (*zhu-cong guanxi* 主從關係) but simply occupied the role of

---

**20** Ibid., "Convention between the United Kingdom and Russia Relating to Persia, Afghanistan, and Tibet," 1907, http://foto.archivalware.co.uk/data/Library2/pdf/1907-TS0034.pdf (accessed April 14, 2019).

**21** Scott Relyea, "Indigenizing International Law in Early Twentieth-Century China: Territorial Sovereignty in the Sino-Tibetan Borderland," *Late Imperial China* 38, no. 2 (2017): 49–50.

**22** Feng, *Zhong-Ying Xizang jiaoshe yu Chuanzang bianqing*, 183–184.

**23** Ibid., 232–233; Lamb, *McMahon Line (I)*, 172–195.

**24** TNA, FO 371/35755, telegram from Mr. Peel (IO) to Mr. Ashley Clarke (FO), May 7, 1943.

**25** Zhu, *Xizang zhuanshi*, 17–22.

a secular patron to a religious priest (*tanyue guanxi*).[26] Boundary conflicts soon broke out between Sichuan and eastern Tibet.

In these circumstances, the British intervened and forced the Chinese republican government into tripartite negotiations at Simla, the summer capital of India, to review the status of Tibet. London returned to the line that "Tibet is under the suzerainty, but not the sovereignty, of China."[27] In the face of British pressure, Chen Yifan 陳貽範, the Chinese representative, gave in. It was tentatively agreed that the Chinese should recognise "the autonomy of Outer Tibet" and "abstain from all interference in the administration of Outer Tibet (including the selection and installation of the Dalai Lama)"; that "Outer Tibet shall not be represented in the Chinese Parliament or any similar body"; and that China should not send troops, officials or colonists into Outer Tibet except for a Chinese amban and a small escort, fewer than 300 men, to reside in Lhasa.[28]

However, the negotiations eventually broke down due to controversies over the physical limits of Tibet. Lhasa's claim for all territories that Tibetan-speaking people inhabited was rebuffed by Beijing as many areas in Kham and Amdo (Qinghai 青海) had been under Chinese rule for centuries.[29] For this reason, a division of Tibet into two parts was proposed by Henry McMahon, the Foreign Secretary of the British-Indian government and the British representative at the conference, in order to solve the problem. Under such a division, Outer Tibet would be an autonomous domain of the Dalai Lama under Chinese suzerainty, while in Inner Tibet, the Chinese would enjoy a degree of sovereignty except for religious issues and some economic affairs falling within the Dalai Lama's sphere.[30] Yet the demarcation between Outer Tibet and Inner Tibet was still unresolved. In addition to demanding that the Qinghaihu 青海湖 (Kokonor) region be placed under its rule, the Chinese final vision of the border ran along the Nujiang 怒江 (the Chinese name of the Salween river in Chinese territories) with some Chinese safeguards over the tract between the Nujiang and Gyamda (Gongbujiangda 工布江達), the western part of the planned Xikang province.[31] Lönchen Shatra (Lunqin Xiazha 倫欽夏箚), the Tibetan representative, however, declined to concede Chamdo (Changdu 昌都). He drew the border along the Jinsha 金沙 (the name for the upper reaches of the Yangtse) and Lancang 瀾滄 (the

---

26 Ibid., 57–58.
27 TNA, FO 371/1613, telegram from Viceroy to Secretary of State (IO), November 12, 1913 and telegram from Hirtzl to Langley, December 2, 1913. Quoted in Lamb, *McMahon Line (II)*, 484.
28 Goldstein, *Demise of the Lamaist State*, 75; Lamb, *McMahon Line (II)*, 495–502, 620–625.
29 Goldstein, *Demise of the Lamaist State*, 68–71.
30 Lamb, *McMahon Line (II)*, 490–491.
31 TNA, FO 535/17, telegram from Chinese Minister to Grey, April 6, 1914. Quoted in ibid., 502.

Chinese name of the Mekong river in Chinese territories) rivers. In the end, the Chinese delegation completely quit the negotiations and the oral agreed political terms were also dropped.[32]

The British and Tibetan stances in the negotiations did not fully overlap. Although a weak and neutral Tibet would benefit the security of British India's northern boundary, London had little interest in supporting the full independence of Tibet as it would create international discord and outrage China and Russia. The best approach, from the British perspective, was for Tibet to remain under symbolic Chinese suzerainty but guarantee its extensive autonomy, including operating a direct diplomatic link with British India.[33] Lhasa was thus pressured to admit Chinese suzerainty. The British, in addition, had further-reaching territorial ambitions regarding Tibet. Unbeknownst to the Chinese, Lönchen Shatra had been induced to agree to a new Indo-Tibetan borderline, later known as the McMahon Line, in parallel with the tripartite Simla Conference negotiations.[34] As a result, the Assam Himalayas or South Tibet (roughly 90,000 square kilometres territories), including Tawang (Dawang 達旺), the birthplace of the Sixth Dalai Lama, were ceded to Britain. Through this concession, the Tibetan delegation had hoped to enlist Britain's endorsement of a Greater Tibet, but it failed to secure any written British assurances. After the breakdown of the conference, the Tibetans naively believed that they were not bound by the McMahon Line on the grounds that the British had failed to manage a new Sino-Tibetan border arrangement to the Tibetans' advantage.[35]

---

32 Interestingly, Chen Yifan, threatened by McMahon with the prospect of recognition of Tibetan independence, had initialled the treaty draft in both political and territorial terms without Beijing's authorisation. Chen was a weak negotiator and was badly prepared for the negotiations. In contrast, Lönchen Shatra, the Tibetan representative, brought cartloads of documents, including the text of the Sino-Tibetan Treaty of 822 A.D., copies of monastery grants and letters of submission by tribal chiefs from districts as far to the east as Tachienlu (Dajianlu 打箭爐, now Kangding 康定). See ibid., 480–481, 505.

33 Goldstein, *Demise of the Lamaist State*, 74.

34 This was achieved through two secret exchanges of notes signed respectively on March 24 and 25, 1914. Yet the British approach apparently conflicts with its previous accords with China in 1906 and Russia in 1907 in which Britain was barred from direct negotiations with Tibet except through the Chinese government. It is also doubtful if the Lhasa government, without any international recognition of its independence or autonomy at the time, had the right to deal with territories that the Chinese government simultaneously claimed.

35 According to the memoir of Lhalu Tsewang Dorje (Lalu Siwangduoji 拉魯·次旺多吉), the McMahon line notes were not ratified by the Thirteenth Dalai Lama, and neither the Dalai Lama nor Lönchen Shatra passed the notes to the Kashag or the National Assembly of Tibet. Indeed, some Kalons had no idea about the exchanges of notes until Charles Bell, the British Political Officer in Sikkim, visited Lhasa in 1920–21. See Lalu Siwangduoji, "Deli mimihuanwen

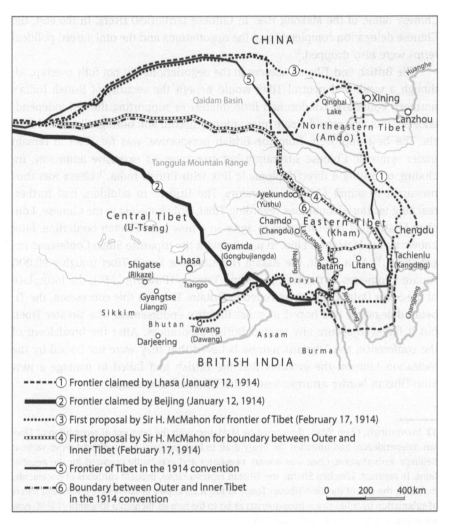

**Map 4:** The Negotiation Process of Boundary Problems between Tibet and China in the Simla Conference.
Based on Ryosuke Kobayashi, "Tibet in the Era of 1911 Revolution," *Journal of Contemporary East Asia Studies* 3, no. 1 (2014): 107. Originally from Singh, *Himalayan Triangle*, map 2.
© Zhaodong Wang & © Peter Palm, Berlin.

weicengdedao yuan Xizang difangzhengfu chengren [The New Dehli Secret Exchanges of Notes Had Never Been Recognised by the Former Local Tibetan Government]," in *Xizang wenshiziliao xuanji (10) [Selected Collections of Historical Sources regarding Tibet (Vol. 10)]*, ed. Xizang zizhi-qu zhengxie wenshiziliaoyanjiuhui (Beijing: Minzu Chubanshe, 1989), 8–9.

On July 3, 1914, after the premature departure of the Chinese from Simla, McMahon and Lönchen Shatra signed a joint Anglo-Tibetan declaration to the effect that the convention would be binding on Britain and Tibet and that China would be barred from the enjoyment of all privileges accruing from the convention unless it also signed.[36] A new Anglo-Tibetan trade agreement was also to be put in place to supersede the Sino-British counterparts of 1893 and 1908. In the end, however, the Simla Conference in 1914 failed to provide either complete independence for Tibet or an agreed Sino-Tibetan boundary. The Kashag merely obtained a British promise of maintaining Tibet's autonomous status.[37] As far as the later views of the Nationalists were concerned, this conference created a sinister precedent and implied that Tibet could enjoy equal negotiating status with China.[38]

### 1.4 The Nationalist Stance towards Tibet, 1927–1941

The political climate in China dramatically changed when Chinese nationalism came to prominence after 1919. As a result, the Chinese claims of sovereignty over Tibet were reinforced. The Republican government was now determined to restore the status quo of the Qing Dynasty's rule over Tibet. In ten principles stipulated by the Chinese Foreign Ministry in 1923, all of Tibet was considered to be Chinese territory, and both external and internal affairs, as well as transportation, were to be placed under the Chinese authorities.[39] Meanwhile, the British official line after 1921 was to recognise Chinese suzerainty over Tibet symbolically but deal with Tibet in practice as a political entity separate from China.[40] With no hope of bringing the Chinese back to the negotiating table, the British govern-

---

36 TBL, IOR/L/P&S/12/4188, "Declaration" enclosed in the telegram from Government of India to India Office, July 23, 1914. The McMahon line was delineated tactfully in the map annexed to the 1914 Simla convention to illustrate its Article 9. It was neither particularly named nor drawn but treated as an extension of Tibet's borderline with China.

37 When Bell visited Lhasa in 1920–21, he indicated that the British had a moral responsibility for safeguarding the interests of Tibet viz-a-viz China. See TBL, IOR/L/P&S/12/4171, letter from Political Officer in Sikkim to Foreign Secretary of Government of India, November 27, 1928.

38 IMH Archives, 11–29–26–00–022, *Feichu Zhong-Ying guanyu Xizang zhi bupingdengtiaoyue* [The Abolition of Unequal Treaties between China and Britain regarding Tibet], the opinion of the MTAC over Tibet during the Sino-British negotiations of the 1943 Treaty, October 31, 1942.

39 Feng, *Zhong-Ying Xizang jiaoshe yu Chuanzang bianqing*, 401–402.

40 TBL, IOR/L/P&S/12/4194, minute by H.A.F. Rumbold, May 22, 1943.

ment declared that the stipulations of the 1914 convention would come into force in September 1925.[41]

The Guomindang regime inherited this strong nationalist position on the Tibetan question. This involved renegotiation. In 1928, Wang Zhengting 王正廷, the Chinese Foreign Minister, sought to revise the 1908 Sino-British trade agreement over Tibet which had included clauses about extraterritorial rights.[42] But the British response was to "ignore the Chinese note, and if it cannot be ignored, to return a short reply without reasons."[43] It also involved redefining the Tibetan question as a domestic rather than a diplomatic issue.[44] Following this logic, the Chinese sought to deal with Britain as an external threat to Tibet while, at the same time, focusing their efforts on Han-Tibetan conciliation.[45]

In the meantime, Tibet witnessed some new developments. The Thirteenth Dalai Lama's effort to establish a centralised executive system over all culturally Tibetan regions achieved some success but strained his relations with outlying monastic polities that wished to be separate and independent. In 1923–1924, the Ninth Panchen Lama and Norlha Hutuktu (Nuona Hutuketu 諾那·呼圖克圖) sought asylum in China proper. Their departure, particularly that of the Panchen Lama, caused anxiety to the Thirteenth Dalai Lama and forced him to ease Lhasa's tensions with Nanjing. The Panchen Lama, only inferior to the Dalai Lama himself, openly advocated that Tibet was part of the Chinese territory and called for the Chinese troops to escort him back to Tibet.[46] Furthermore, the Thirteenth Dalai Lama's modernising reforms, principally in military areas, provoked serious discontent among monasteries in Lhasa.[47] As a result, British elements behind the reform were blamed while pro-Chinese sentiments gradual-

---

**41** TBL, IOR/L/P&S/12/4171, minute paper by J.C. Walton, November 22, 1928.

**42** According to terms of itself, the 1908 convention could be revised every ten years. TBL, IOR/L/P&S/12/4171, note from Chinese Foreign Ministry to Foreign Office, October 9, 1928; UK Treaties Online, "Regulations respecting Trade in Tibet, concluded between the United Kingdom, China and Tibet," 1908, http://foto.archivalware.co.uk/data/Library2/pdf/1908-TS0035.pdf (accessed April 13, 2019).

**43** TBL, IOR/L/P&S/12/4171, minute paper by J.C. Walton, November 22 1928.

**44** IMH Archives, 11–01–17–01–01–027, *Xizang wenti (jiejue Zhong-Zang yijianshu)* [The Tibetan Question (Proposals for Solving the Sino-Tibetan Question)], selected opinions of French consultant, Georges Padoux (Baodao 寶道 as his Chinese name) on the Tibet question, unspecific date; and telegram from Chiang Kai-shek to Luo Wengan (Foreign Minister), December 5, 1932.

**45** IMH Archives, 11–01–17–01–01–027, *Xizang wenti (jiejue Zhong-Zang yijianshu)* [The Tibetan Question (Proposals for Solving the Sino-Tibetan Question)], report from Luo Wengan 羅文幹 to Chiang Kai-shek and Wang Jingwei, July 31, 1933.

**46** Smith Jr., *Tibetan Nation*, 216–219.

**47** Goldstein, *Demise of the Lamaist State*, 89–110.

ly grew in the religious sector, thanks in part to the Nationalist promotion of Tibetan Buddhism in China proper and generous Nationalist patronage to Tibetan monasteries.[48] Partly because of this, the amicable Britain-Tibet relationship gradually cooled in the late 1920s. Other factors also contributed to the result, such as British-Tibetan disputes over trade and territorial issues, but the crucial one was Britain's failure to make China accept the Simla accord. All these developments gave Nanjing some leverage to intervene in Tibetan politics. In 1933, capitalising on the death of the Thirteenth Dalai Lama, the Chinese established an official residency at Lhasa.[49] The Fourteenth Dalai Lama hailed from Amdo and thus from the Chinese-dominated area. The Kashag was thus obliged to invite Chinese representatives to attend the enthronement ceremony in 1940. The Nationalist regime duly interpreted this as a resumption of Chinese traditional rights in presiding over the ceremony. Moreover, the Regent Radreng Hutuktu (Rezhen Hutuketu 熱振·呼圖克圖), who held power in Tibet until the Fourteenth Dalai Lama was old enough to assume political duties, had some pro-Chinese sentiments.

Yet the Nationalist government, which was preoccupied with other urgent domestic issues, did not have the political intent or the military strength to reconquer Tibet in the 1930s. In fact, the Guomindang-controlled domain was never directly adjacent to any parts of Tibetan-controlled territories. For most of the 1930s and 1940s, the border regions with Tibet were occupied by two major semi-independent warlords, Liu Wenhui 劉文輝 and Ma Bufang 馬步芳. Liu began his rule over the eastern areas of Kham in 1931–32. Later, he was appointed by the Nationalist government as governor of Xikang province, which incorporated western Sichuan and Eastern Tibet.[50] Liu was eager to fight against the Tibetans because most parts of his Xikang province were, in practice, under Tibetan authority. However, Chiang Kai-shek had no wish to finance a warlord with questionable loyalty to him.[51] Meanwhile, Amdo was ruled by a branch of a Chinese Muslim family named Ma to which Ma Bufang belonged. The rule of this family over Northwest China had been formed in the 1870s, when Zuo Zongtang 左宗棠 reoccupied Xinjiang, and had endured since this period. Nonetheless, despite these constraints, Tibet was highly symbolic for the Nationalists and was thus incorporated as part of the Chinese territory in the Guomindang

---

**48** Tuttle, *Tibetan Buddhists in the Making of Modern China*, 156–192.
**49** The British soon acquired the same right to station a British mission at Lhasa as a counter to increased Chinese influence.
**50** Xikang was one of six new provinces established in the 1930s in China's outlying territories. The others were Chaha'er 察哈爾, Rehe 熱河, Suiyuan 綏遠, Ningxia 寧夏 and Qinghai 青海.
**51** Lin, *Tibet and Nationalist China's Frontier*, 43–44.

Manifesto of 1935 and the Draft Constitution of 1936.[52] A Mongolian and Tibetan Affairs Commission (MTAC) was duly established to deal with Tibetan issues, as well as Outer Mongolia.[53] But without substantial control over the aforementioned territories, the MTAC was more akin to an advisory and planning group to rebuild Chinese affinity with Outer Mongolia and Tibet in the hope that the latter might offer symbolic subordination to Chinese authority.

The Second World War, however, changed the picture. The geographic distance between Nationalist domain and Tibet was considerably shortened as the Nationalist influence was driven to Southwest China by the Japanese advance. The Nationalists were for the first time able to exert more direct force over Tibet. But Sino-Tibetan relations again soured when Taktra Rimpoche (Dazha Huofo 達紮活佛) replaced Radreng Hutuktu as Regent in 1941. Compared to his predecessor, Taktra Rimpoche had a stronger desire for Tibetan independence and thus was more alert to increasing Chinese influence. His rule covered the period 1941–1950 and it was against this background that the Sino-British wartime confrontation over Tibet ensued.

## 2 Tibet as a Supply Route to Support Wartime China

The start of the Pacific War complicated the Sino-Tibet-British dynamics. Britain and China, the long-time rivals regarding Tibet ever since the late nineteenth century, were now allies in the same camp. More importantly, Tibet somewhat unexpectedly came to the Allies' attention as a potential passage to supply China after the Japanese blockaded all Chinese access to the outside world with the exception of some channels to the Soviet Union. These new situations created some advantages to help the Nationalists advance their claim to sovereignty over Tibet but also posed a challenge to Britain's traditional policy of maintaining Tibet's *de facto* independence from Chinese interference.

### 2.1 Building a Vehicle Road between British-India and Southwest China

The Nationalist regime was on the verge of bankruptcy before the start of the Pacific War. As China was predominantly an agricultural country, its war effort heavily depended upon import of arms and munitions from outside. This became

---

**52** Lamb, *Tibet, China & India, 1914–1950*, 334.
**53** Lin, *Tibet and Nationalist China's Frontier*, 33–34.

extremely difficult after the Japanese occupied all Chinese seaports. The situation further deteriorated in the summer of 1940. Under Japanese pressure, Britain temporarily closed the Burma Road, the overland supply route from Burma to Southwest China, which coincided with the Japanese invasion of northern French Indochina and the blockade of the rail line between Hanoi and Kunming. Though the Burma Road reopened in October 1940, Chiang Kai-shek was convinced that to transport over 80% of Chinese international supplies along this single route was too risky.[54] He therefore urgently looked for alternative means of connecting Indian Assam and Chinese Yunnan.

The Foreign Office and the British ambassador to Chongqing, Archibald Clark Kerr, welcomed this aspiration. Clark Kerr recommended a vehicle route based on air examination, starting from Sadiya in India to Tachienlu (Kangding 康定) by way of Rima (Xiachayu 下察隅 town), Chamdo and Batang 巴塘.[55] Yet the Chinese preferred a more convenient one (dubbed the Rima road), starting from Sadiya to Kunming via Rima, Menkong (Mangkang 芒康), Atuntze (Deqin 德欽) and Dali 大理, where it met the old Burma road.[56] Chongqing favoured this route also because a road passing through the Kham areas would aid Chinese penetration into eastern Tibet and avoid the disputed Sino-Burmese boundary.[57] Clark Kerr agreed to the route owing to war exigencies (see map 5).[58]

However, the India Office and the British-Indian government disapproved of these proposals. They were alerted by Chinese ambitions in Tibet and the Assam Himalayas (the McMahon area).[59] The Foreign Office, by contrast, believed that the war against the Japanese outweighed any British interests in these areas. A small party of Chinese surveyors and engineers was thus allowed to examine the feasibility of their route.[60] In this context, the India bureaucrats sought to exploit Tibet's antagonism towards the Chinese.

Before the British contact, the Kashag was ambivalent over the matter. It desired to refuse the road scheme but feared the resultant Chinese punishment. New Delhi's attitudes strengthened its objection. It had been reported by Kong Qingzong 孔慶宗, the head of the Chinese mission at Lhasa, that the Kashag

---

**54** Lamb, *Tibet, China & India, 1914–1950*, 309.
**55** Ibid., 305.
**56** Ibid.
**57** IMH Archives, 11–29–19–06–020, *Zang-Yin yiyunxian 1* [Indo-Tibet Pack Route Line (Part 1)], letter from Foreign Ministry to the Investigation and Statistics Bureau of the Military Affairs Commission (ISBMAC), June 11, 1942.
**58** Lamb, *Tibet, China & India, 1914–1950*, 305–306.
**59** Guyot-Réchard, *Shadow States*, 63.
**60** TBL, IOR/L/P&S/12/4613, telegram from Chongqing to Foreign Office, April 22, 1941.

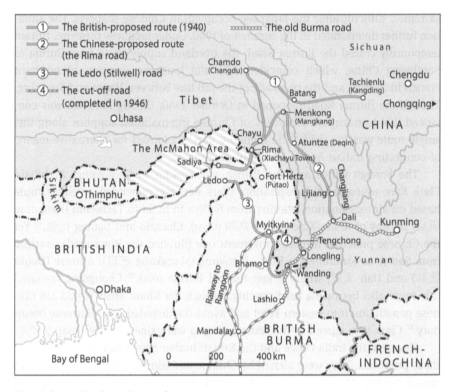

**Map 5:** Respective Route Proposals.
© Zhaodong Wang & © Peter Palm, Berlin.

had agreed to a ground survey of the proposed Rima road.[61] But the Kashag later denied any acceptance of road construction in Tibetan territories and stated the refusal had been affirmed by the Tsongdu (*minzhong dahui* 民眾大會, Tibetan National Assembly).[62] Lhasa then expelled the unarmed Chinese surveyors and also blew up bridges and roads on the border in order to prevent Chinese re-entry.[63] Chinese officials in Chongqing duly suspected that behind the scenes

**61** TBL, IOR/L/P&S/12/4613, telegram from Clark Kerr to Foreign Office, September 15, 1941; Chen, *Kangzhanqianhou zhi Zhong-Ying Xizang jiaoshe*, 144.
**62** TBL, IOR/L/P&S/12/4613, telegram from the Indian government to India Office, July 25, 1941; telegram from the Indian government to India Office, September 29, 1941; and telegram from the Indian government to India Office, September 30, 1941.
**63** Lin, *Tibet and Nationalist China's Frontier*, 125.

were the British playing a trick.[64] As a result, only an Anglo-Chinese aerial survey was permitted.[65] Back in London, the Foreign Office also began to have second thoughts. It was now convinced that the construction of the Rima road would be fraught with terrain difficulties and unable to greatly improve China's military effort. The Foreign Office thus withdrew its support and informed the Chinese of British non-participation "unless the full and willing assent of the Tibetan Government was forthcoming."[66]

To placate Chiang Kai-shek, New Delhi suggested a "Southern Line" instead, a vehicle route which would run through the extreme north of Burma and avoid any Tibetan territories.[67] Although this plan ran the risk of impairing British claims over Northern Burma, it was deemed better than jeopardising the British "treaty" rights over the hinterland south of the McMahon Line.[68] London then ratified the proposal and recommended it to Chiang. The vehicle road (the Ledo Road) would stretch from Ledo in India and enter Yunnan by the way of Indian Assam via either Fort Hertz (Putao) or Myitkyina in northern Burma (see map 5).[69] This overland route was accepted by Chiang in February 1942 when he undertook a short visit to India.[70] However, the subsequent fall of Myitkyina to the Japanese in May 1942 halted the plan. It was not until January 1945, after the recovery of Myitkyina, that the Ledo Road finally materialised. It became known as the "Stilwell Road".

Despite this, the Chinese government had never fully given up the Rima highway scheme. In early 1944, another survey party, masquerading as merchants, was sent to eastern Tibet. They departed from Kunming (the capital of Yunnan) and infiltrated into the Lhasa-administered tribal areas in the Assam-Tibetan borderlands, eventually arriving in Sadiya. Yet due to Tibetan and British opposition, as well as severe practical difficulties, the project was never put into practice. The part of the route that lay in the Chinese territories was, however, subsequently completed by the People's Republic of China (PRC) in the late 1950s.[71]

---

64 IMH Archives, 11–29–19–06–020, *Zang-Yin yiyunxian 1* [Indo-Tibet Pack Route Line (Part 1)], letter from Foreign Ministry to ISBMAC, June 11, 1942.
65 TBL, IOR/L/P&S/12/4613, telegram from the Indian government to India Office, May 12, 1941.
66 TBL, IOR/L/P&S/12/4613, telegram from Foreign Office to Chongqing, October 31, 1941.
67 TBL, IOR/L/P&S/12/4613, telegram from the Indian government to India Office, May 12, 1941.
68 Lamb, *Tibet, China & India, 1914–1950*, 307.
69 TBL, IOR/L/P&S/12/4613, telegram from Foreign Office to Chongqing, January 21, 1942.
70 February 14, 1942, see Meihua Zhou, ed. *Jiang Zhongzheng zongtong dang'an: shiliie gaoben [Draft Papers of Chiang Kai-shek]*, 82 vols., vol. 48 (Taipei: Academia Historica, 2011), 304–305.
71 Lin, *Tibet and Nationalist China's Frontier*, 135–136.

## 2.2 Utilising the Existing Pack Routes in Tibet

The Anglo-American strategy of "Europe First" rendered China's persistent fight against Japan extremely vital in East Asia, especially in the first years of the Pacific War. Given that the capacity for air transportation over the "Hump" was still limited,[72] pressure was consistently imposed on Britain to find other overland trans-Tibet routes to support China. Horace Seymour, the British ambassador to Chongqing who succeeded Clark Kerr, stressed that "irrespective of the practical importance of the route ... the Chinese must be given hopes of a supply route being opened."[73] From the British perspective, China's persistence also contributed to the defence of British India, the core of the British Empire in Asia.

Against this background, the British sought to facilitate a Tibetan animal pack route, which would have a smaller transport capacity but might be deemed less of a threat to Tibet's autonomy. However, the Kashag initially refused this proposal.[74] The main reason was its anxiety over getting embroiled in the war.[75] At this time it had no confidence that the Allied nations would triumph. Indeed, an atmosphere prevailed among high-ranking Tibetan officials that final victory might well belong to the Axis powers and that the Japanese would eventually conquer all of China.[76] But the British were determined to exert pressure through their Mission at Lhasa, and Frank Ludlow, the head of the Mission, used "a judicious blend of bribes and threats of economic sanctions" to force the Regent's hand.[77] In this case, any sensitivities over Tibet gave way to the need for cooperating with the Chinese. It was believed by London that "something must be done even though the practical results in the way of supplies will be small, and it will probably involve the collapse of our valuable Tibetan policy."[78] The suggested pack route was to follow the traditional

---

72 US pilots were specifically instructed to avoid Tibetan airspace. See IMH Archives, 11–29–19–06–020, *Zang-Yin yiyunxian 1* [Indo-Tibet Pack Route Line (Part 1)], letter from Foreign Ministry to ISBMAC, June 11, 1942.

73 TBL, IOR/L/P&S/12/4614, telegram from Chongqing to Foreign Office, May 22, 1942.

74 TBL, IOR/L/P&S/12/4613, telegram from Foreign Office to Washington, May 15, 1942.

75 IMH Archives, 11–29–19–06–020, *Zang-Yin yiyunxian 1* [Indo-Tibet Pack Route Line (Part 1)], letter from E. Teichman to Long Liang, July 9, 1942.

76 Melvyn C. Goldstein, Dawei Sherap and William R. Siebenschuh, *A Tibetan Revolutionary: the Political Life and Times of Bapa Phüntso Wangye* (Berkeley, Los Angeles and London: University of California Press, 2004), 77–78.

77 According to Alastair Lamb's research, this was the Regent's own decision without consulting the Tsongdu. See Lamb, *Tibet, China & India, 1914–1950*, 310.

78 TBL, IOR/L/P&S/12/4614, minute by R. Peel (IO), May 23, 1942.

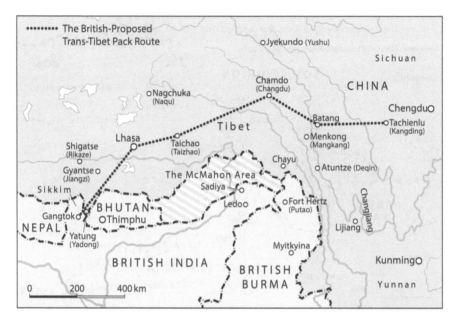

**Map 6:** The British-proposed Trans-Tibet Pack Route.
© Zhaodong Wang & © Peter Palm, Berlin.

trade route, starting from Sikkim to Lhasa and then proceeding to Batang and Tachienlu by way of Chamdo (see map 6).

Besides this pack route, the British also endeavoured to collaborate with China to open up two other supply lines. The first was the Gilgit-Hunza-Kashgar (Kashi 喀什) pack route from British Kashmir to Chinese Xinjiang via the Pamir plateau (see map 7).[79] The capacity of this caravan route was also very small, but it was important symbolically as it was "one of the few purely British measures ... to assist the Chinese war effort."[80] The second was a long-distance line comprised of both a railway and a vehicle road, the Zahedan-Meshed (Mashhad)-Alma Ata (Almaty)-Lanzhou 蘭州 line (see map 7). The line ran from India to China through Eastern Persia, Russian Central Asia and Chinese Xinjiang.[81] The Russians were reluctant to initiate any transit via Russian Turkestan al-

---

[79] About trans-Pamir supply routes, there was an alternative to the route, the Leh-Karakoram pack route. But the Chinese government did not prefer it. See TNA, CAB 111/336, letter from Sir H. Seymour to Government of India, August 10, 1942.
[80] TNA, CAB 111/336, telegram from R. Peel (IO) to O.K. Caroe (New Delhi), February 4, 1943.
[81] Ibid; TNA, CAB 111/336, extract from Chongqing monthly summary for November 1942, December 7, 1942.

**Map 7:** The British-proposed Gilgit-Hunza-Kashgar Pack Route and the Zahedan-Meshed-Alma Ata-Lanzhou Line.
© Zhaodong Wang & © Peter Palm, Berlin.

though they had agreed with Washington.[82] At this time, the Soviet Union had no wish to infuriate Japan considering the latter's serious military threat to Southeast Siberia.

These British initiatives confused the Nationalists given their difficult relationship over Tibet.[83] They were naturally suspicious of British motives. Some Chinese officials even speculated that British intentions were not to facilitate trans-Tibetan supplies but instead secure a line of retreat if India fell to the Japanese.[84] But Chiang Kai-shek took a more pragmatic stance. He saw this rapprochement as a great opportunity to increase Chinese influence in the hinter-

**82** TNA, CAB 111/336, document of supply routes to China (views in the U.S. War Department), December 31, 1942.

**83** IMH Archives, 11–29–19–06–020, *Zang-Yin yiyunxian 1* [Indo-Tibet Pack Route Line (Part 1)], letter from Foreign Ministry to ISBMAC, May 14, 1942.

**84** IMH Archives, 11–29–19–06–020, *Zang-Yin yiyunxian 1* [Indo-Tibet Pack Route Line (Part 1)], letter from Foreign Ministry to ISBMAC, June 11, 1942.

land of Tibet.[85] To this end, the Chinese Ministry of Communications proposed a counter plan for the proposed route from Sikkim. The Guomindang government would supervise parts of the route in the Tibetan domain, assigning Chinese liaison officers to reside at intermediate stations, such as Yatung (Yadong 亞東), Gyantse (Jiangzi 江孜), Lhasa, Taichao (Taizhao 太昭) and Chamdo. Radios would be installed in these places.[86] The Chinese emphasised that these officials "shall be given strict instructions not to engage themselves in matters other than their own duties to supervise transportation."[87] Yet, through their presence, the Chinese could embed themselves across the Tibetan plateau. More importantly, a right of supervision implied Tibet's subordinate status to the Chinese central government.

The Kashag was clearly aware of Chinese ambitions and refused such an arrangement. Its view was that Lhasa and Chamdo should be excluded from any Tibetan pack route since Lhasa was the capital while Chamdo was the military and political centre of eastern Tibet. Furthermore, the Kashag argued that neither Chinese nor British officials should accompany goods through Tibet or supervise the transportation[88] and that only non-military supplies, such as medical stores, lubricating oils and mail, could be transported across Tibet.[89] Lhasa also insisted on using British India as an intermediary for these negotiations.[90] In July 1942 it even established a foreign affairs body, the Tibetan Foreign Bureau (TFB), to deal with these issues. As a result, the opening of the pack route was delayed.[91]

The British sided with the Tibetans. They advised the Chinese government to "give [a] definite and public undertaking of intention to respect Tibetan autono-

---

85 This is different from Lin Hsiao-ting's argument, that the reason for Chiang Kai-shek accepting the British proposal was purely to obtain international supplies. See Lin, *Tibet and Nationalist China's Frontier*, 126 – 127. In fact, Chiang's main aim was to use the route to expand Chinese influence in Tibet.

86 IMH Archives, 11–29–19–06–020, *Zang-Yin yiyunxian 1* [Indo-Tibet Pack Route Line (Part 1)], letter from the Ministry of Communications to MTAC, June 30, 1942.

87 IMH Archives, 11–29–19–06–020, *Zang-Yin yiyunxian 1* [Indo-Tibet Pack Route Line (Part 1)], letter from Long Liang to Eric Teichman, July 7, 1942.

88 IMH Archives, 11–29–19–06–021, *Zang-Yin yiyunxian 2* [Indo-Tibet Pack Route Line (Part 2)], minute of transport of Chinese supplies from India to China via Tibet, August 6, 1942.

89 IMH Archives, 11–29–19–06–020, *Zang-Yin yiyunxian 1* [Indo-Tibet Pack Route Line (Part 1)], letter from Eric Teichman to Long Liang, July 9, 1942. According to Alastair Lamb's research, petroleum products eventually received Tibetan consent. See Lamb, *Tibet, China & India, 1914 – 1950*, 310.

90 TNA, CAB 111/336, telegram from Sir H. Seymour to Foreign Office, December 22, 1942.

91 Chongqing refused to recognise the TFB. As a result, Sino-Tibetan talks were conducted through the Tibetan representatives in Chongqing. See Chen, *Kangzhanqianhou zhi Zhong-Ying Xizang jiaoshe*, 146 – 147.

my and to refrain from interfering in Tibet's internal administration" so as to dispel the Tibetans' anxieties.[92] However, Fu Bingchang, the Chinese Vice Foreign Minister, refused to do this by claiming that

> there was no occasion for giving assurances regarding 'autonomy'; that Tibet was considered a part of the Republic of China; but that China had no intention of altering the situation whereby internal administration in Tibet is in fact autonomous. [93]

The British then sought a tripartite agreement to ratify the route, but the Chinese refused this.[94] Any such agreement would indicate that Britain enjoyed an equal political status with China in Tibet.

In an aide-memoire to the US Department of State, London laid the blame for the impasse on aggressive Chinese ambitions over Tibet:

> the initiative for the organisation for a pack route from Kalimpong via Central Tibet to China was taken by the Government of India. ... [When] the time came to work out practical details the Chinese Government made certain stipulations in regard to supervision of this route by Chinese officials, stipulations which the Tibetan Government were unable to accept. The Chinese Government moreover opposed any form of tripartite agreement in which the British Government would participate. ... Lack of further progress had been due to the unforthcoming attitude of the Chinese and to the Tibetan Government's suspicion of Chinese intentions.[95]

Lin Hsiao-ting argues that the Chinese had agreed to the tripartite approach for pragmatic reasons and inferred from this decision that Tibet's place within Nationalist China was ambivalent during wartime.[96] However, this interpretation is not supported by the Chinese minutes of the joint meeting of Secretariat of Executive Yuan, Foreign Ministry, Finance Ministry, Ministry of Communications, the MTAC and Military Affairs Commission (MAC) of the Nationalist government, etc.[97] These minutes suggest that the Chinese refusal derived from practical as well as political concerns. He Yingqin 何應欽, the Chief of Staff of the MAC, criti-

---

**92** Telegram from the Secretary of State to the Ambassador in China, July 3, 1942, US Department of State, ed., *FRUS: 1942 China*, 626 (hereafter *FRUS: 1942 China*).

**93** Telegram from Ambassador in China (Gauss) to the Secretary of State, July 13, 1942, ibid., 627.

**94** Telegram from *Chargé d'affaires* in India to the Secretary of State, May 15, 1943. Ibid., ed., *FRUS: 1943 China*, 631 (hereafter *FRUS: 1943 China*).

**95** Aide-memoire, from the British Embassy to the US Department of State, April 19, 1943. Ibid., 626–628.

**96** Lin, *Tibet and Nationalist China's Frontier*, 127.

**97** IMH Archives, 11–29–19–06–022, *Zang-Yin yiyunxian 3* [Indo-Tibet Pack Route Line (Part 3)], minutes of the Joint Meeting of the Executive Yuan, Foreign Ministry, Finance Ministry, Ministry of Communications, the MTAC and the MAC, etc., September 15, 1942.

cised the lack of efficacy of the pack route. In a report to Chiang Kai-shek, he deemed the route "worthless" if it did not help Chinese claims over Tibet because it might cost over three million silver Rupees to operate per year but convey at most 2,000 tons of non-military goods.[98] He advised Chiang to either abandon the project or circumvent the Kashag to do local deals with Tibetan traders and carriers as a first step towards securing China's goals in Tibet. Chiang chose the latter.[99] This plan was communicated to New Delhi and soon put into practice.[100]

Lhasa acknowledged the proposal through British channels. It had no objection to deal with the matter on a purely commercial basis as long as British India acted as an intermediary. However, the British did not pass this opinion on to Chongqing due to their worries that this might reopen the discussion over the political status of Tibet. They decided to wait and see how far the Chinese could get in negotiating commercial contracts with Tibetan traders and carriers at Kalimpong.[101] Chinese unilateral contacts, however, infuriated Lhasa, which forbade Tibetan traders and carriers from carrying Chinese goods without its approval.[102] When Chongqing approached the Kashag over the issue in January 1943, it was told that Tibetan permission would not be forthcoming unless Britain was a third party to the scheme.[103] The Chinese again objected to any form of a tripartite accord.[104] They suspected the British of being the manipulators behind Tibet's intransigent attitude.[105]

---

98 According to American calculations, the capacity was 1,000 tons. See telegram from Personal Representative of President Roosevelt in India to the Secretary of State, January 26, 1943. US Department of State, ed., *FRUS: 1943 China*, 620–621.

99 IMH Archives, 11–29–19–06–022, *Zang-Yin yiyunxian 3* [Indo-Tibet Pack Route Line (Part 3)], report from He Yinqin to Chiang Kai-shek, September 20, 1942.

100 IMH Archives, 11–29–19–06–022, *Zang-Yin yiyunxian 3* [Indo-Tibet Pack Route Line (Part 3)], telegram from Foreign Ministry to the Chinese commissioner at India, October 12, 1942.

101 TBL, IOR/L/P&S/12/4210, minute by H.A.F. Rumbold, April 22, 1943.

102 IMH Archives, 11–29–19–06–022, *Zang-Yin Yyiyunxian 3* [Indo-Tibet Pack Route Line (Part 3)], letter from the MTAC to Foreign Ministry, February 24, 1943.

103 TBL, IOR/L/P&S/12/4210, minute by H.A.F. Rumbold, April 22, 1943.

104 IMH Archives, 11–29–19–06–022, *Zang-Yin yiyunxian 3* [Indo-Tibet Pack Route Line (Part 3)], letter from Foreign Ministry to the MTAC, Ministry of Military Ordinance, Ministry of Transport, and the Transport Conference Committee, June 11, 1943.

105 Telegram from Personal Representative of President Roosevelt in India to the Secretary of State, January 26, 1943. US Department of State, ed., *FRUS: 1943 China*, 620–621.

## 3 Sino-Tibetan Military Tensions and Subsequent Sino-British Disputes

In March 1943 the Kashag held up all goods in transit to China, including those being transported by private Chinese traders, pending a formal tripartite settlement.[106] To understand the significance of this, it should be pointed out that the Tibetan pack route had been already developed by Tibetan and Yunnan traders and carriers on an impressive scale.[107] The trade route ran from Kalimpong, at the head of the Chumbi Valley (Chunpihegu 春丕河谷), via Lhasa to the terminus at Lijiang 麗江 on the Jinsha river. The whole journey took 15 weeks. By early 1943 over 5,000 mules were at work, delivering a standard load of 150 lbs (68 kg) per mule and each costing 160 silver Rupees.[108] In 1944–45, the number of mules peaked at 8,000.[109] This route made an important contribution to the wartime prosperity of Lijiang and also explains why the Chinese government was reluctant to accept interference from the Kashag or Britain.

Simultaneously, a rumour about a possible Chinese invasion of Tibet began to spread. This encouraged British officials at the time, as well as some later historians, such as Melvyn C. Goldstein,[110] to link the two events and believe that Chiang Kai-shek was infuriated by the Tibetans' blockade and thus resorted to military action.[111] In fact, Chinese military mobilisation had begun earlier than March 1943. According to Chinese sources, in late 1942 Chiang instructed three military commanders along the Sino-Tibetan border, Ma Bufang, the commander and chairman of Qinghai province, Liu Wenhui, the commander and chairman of Xikang province, and Long Yun 龍雲, the commander and chairman of Yunnan

---

**106** TNA, CAB 111/336, extract from telegram from Chongqing to Foreign Office, April 5, 1943; TBL, IOR/L/P&S/12/4210, telegram from External Affairs Department (GOI) to India Office, April 23, 1943.
**107** TNA, CAB 111/336, letter from Governor General (New Delhi) to India Office, January 4, 1943.
**108** TNA, CAB 111/336, extracts from Persian Gulf War Trade Bureau, April 19, 1943; TNA, CAB 111/336, extract from dispatch No. 75 from Mr Consul-General Ogden to Seymour, April 20, 1943.
**109** Lamb, *Tibet, China & India, 1914–1950*, 311–312.
**110** Melvyn C. Goldstein argues that the Tibetan embargo gave rise to Chiang's military action. See Goldstein, *Demise of the Lamaist State*, 388. Other scholars' works are more obscure on the question, Alastair Lamb and Lin Hsiao-ting for instance. See Lin, *Tibet and Nationalist China's Frontier*, 132–133; Lamb, *Tibet, China & India, 1914–1950*, 316–317.
**111** Some officials of the British-Indian government held the same judgement at the time. See TNA, FO 371/35755 and TBL, IOR/L/P&S/12/4210, telegram from External Affairs Department, Government of India to India Office, April 23, 1943.

province, to move their forces close to the border.[112] But only Ma Bufang actually complied with the order and mobilised his 10,000 Chinese Muslim troops.[113]

Chiang's thinking behind these troop movements merits further consideration. Liu Wenhui and his associates believed that Chiang's target was to weaken the three border warlords, particularly Liu himself. This argument is mainly supported by the reminiscences of Liu's Chief of Staff of the 24th Army, Wu Peiying 伍培英. In his view, two divisions of central government troops were to help Liu defend his territories so that all his troops would be able to join the battle with the Tibetans.[114] As a result, Liu would lose his base in Xikang and deplete his troops in the Tibetan conflict. Alastair Lamb and Lin Hsiao-ting advance a different theory. They maintain that Chiang's underlying aim was to neutralise Japanese penetration in Tibet.[115] There were indeed some unconfirmed rumours at the time in Chongqing about the activities of Japanese monks in Tibet and the preparation of an airfield for the Japan's use. Yet a more convincing explanation seems contained in Chiang's own words. In a talk with the Tibetan representative at Chongqing in May 1943, he remarked that he wished to prevent a possible Tibetan conspiracy with the Japanese as well as protect Sino-Indian road construction and the Tibetan pack route. He also warned that if any collusion with Japan was discovered Tibet would be bombed.[116] It is highly unlikely that he wished to conquer Tibet by force. Chiang only deployed poorly equipped armies in the peripheral areas and if he had really wanted to secure a victory, he would have despatched much better trained and equipped forces. But if the warlord in question was weakened this would be a welcome by-product of the military operations.

The Chinese military deployment adjacent to the Sino-Tibetan border naturally caused much anxiety to the Kashag, exacerbated by the presence of the Nationalist authorities in Xinjiang. Thereafter, the Chinese threat was to a large degree exaggerated by Tibet. It was alleged that eleven Chinese divisions were about to invade. Yet, according to Song Ziwen, "there were scarcely more than

---

**112** Peiying Wu, "Jiang Jieshi jiazheng Zang yitu Kang de jingguo [The Historical Process of Chiang Kai-shek's Manoeuvre of Controlling Xikang on the Pretext of the Tibetan Affairs]," in *Wenshi Ziliao Xuanji (Vol. 33) [Selections of Historical Materials (Vol. 33)]*, ed. Zhengxie Quanguo Weiyuanhui (Beijing: Wenshi Ziliao Chubanshe, 1985), 140. Another Chinese source indicates the timing was mid-1942. See Chiang Kai-shek to Xu Yongchang (Minister of Military Ordinance), October 19, 1942, CB, 09–1413, quoted in Lin, *Tibet and Nationalist China's Frontier*, 133–134.
**113** TBL, IOR/L/P&S/12/4210, telegram from Chongqing to Government of India, April 22, 1943.
**114** Wu, "Jiang Jieshi jiazheng Zang yitu Kang de jingguo," 146–147.
**115** Lin, *Tibet and Nationalist China's Frontier*, 133–135; Lamb, *Tibet, China & India, 1914–1950*, 316–317.
**116** Telegram from Zhou Kuntian 周昆田 (MTAC) to Kong Qingzong 孔慶宗 (Lhasa), May 13, 1943. Quoted in Chen, *Kangzhanqianhou zhi Zhong-Ying Xizang jiaoshe*, 156–157.

two Chinese battalions on the Tibetan border."[117] Seymour even suspected that the rumours of Chinese aggression were deliberately exaggerated by the Tibetans in the hope of stirring up the British-Indian government into taking action on their behalf.[118] Regardless of how serious the Chinese threat was, the Tibetans were determined to fight should China encroach on their territory. In tandem with this, a plea for arms and ammunition was sent to New Delhi. This included no fewer than 16,000,000 rounds of .303 ammunition, 36,000 shells and 20 machine guns, either Lewis or Bren (preferably the latter).[119]

Tibet's request for munitions placed the British in a dilemma. They were forced to reconsider their commitments to Tibet in light of the Anglo-Chinese alliance. In a letter to enlist British support, R.D. Ringang, a Tibetan noble educated at Rugby School who acted as an English interpreter to the Kashag, implied that Britain was Tibet's ally.[120] This concerned the British because, as Major George Sherriff, who took over the Mission at Lhasa from Ludlow in 1943, argued: "it was impossible for Great Britain to be an ally to both China and Tibet."[121] Lhasa was gently reminded that "no formal alliance exists between [the] British government and Tibet ... [but] comment on the incorrect use of the word "ally" in no way implies withdrawal from assurances given to the Tibetan government in the past."[122]

Another thorny question was whether it was appropriate to provide military supplies to Tibet at such a time. Olaf K. Caroe, the Secretary of the External Affairs Department in the Indian government, cautioned that

> it would be a mistake of policy to provide munitions of war to Tibet at this juncture and would lay H. M. G. and Government of India open to propaganda that [the] war effort is being diverted to enable Tibet to resist China. ... We realise that refusal which would of course be based on [British] war needs will leave Tibetans without ammunition to resist aggression should it come. But we should prefer making considerable supply of munitions

---

117 Notes of a Conversation between Dr. Soong and Mr. Churchill, July 29, 1943. See Wu and Kuo, eds., *T. V. Soong: Selected Minutes of Meetings with Foreign Leaders, 1940–1949*, 378.

118 TBL, IOR/L/P&S/12/4194, extract from despatch No. 649 from Seymour to Foreign Office, July 5, 1943.

119 TNA, FO 371/35755, telegram from External Affairs Department (GOI) to India Office, May 1, 1943; Lamb, *Tibet, China & India, 1914–1950*, 317.

120 The translated version of the Tibetan letter can be found in TNA, FO 371/35755, telegram from External Affairs Department (GOI) to India Office, May 1, 1943.

121 These were Sherriff's words quoted by R.D. Ringang. See TBL, IOR/L/P&S/12/4210, letter from R.D. Ringang to Major Sherriff, May 10, 1943. The word "ally" came from a loose translation. More accurate would be "one from whom it is hoped help will be given". See TBL, IOR/L/P&S/12/4210, telegram from Sherriff (Lhasa) to Gould (Gangtok), May 12, 1943.

122 TBL, IOR/L/P&S/12/4210, telegram from New Delhi to Gould (Calcutta), May 7, 1943.

available to Tibet at the close of the war [due] to risk of propaganda aforesaid, relying in meantime on diplomatic support.[123]

As a result, British support was officially limited to diplomatic gestures.[124] But a supply of ammunition (perhaps 5,000,000 rounds of .303) did reach Lhasa in 1944. Some technical assistance was also provided to the Tibetans to help them overhaul the Trapchi arsenal and maintain their mountain guns.[125]

British diplomacy rested on three approaches. The first was to eliminate any possible pretexts for Chinese military aggression. To this end, Ludlow remarked to the Regent that it was unwise to blockade Chinese goods at the present juncture as this would "give [Chongqing] just cause for complaint."[126] Tibetans were thus advised to end the embargo in order to relieve the tension.[127] The pack route was consequently reopened for private carriers under the condition that no military stores were carried or there was no foreign supervision of shipments.[128] In tandem with this, the British managed to obtain two clarifications from the Kashag. One was that Tibet, "being a country entirely devoted to religion ... emphatically [denies] having any dealings or understanding with other foreign powers [referring to Japan]."[129] The other was that the concentration of the Tibetan troops on the Qinghai-Tibetan border was "purely for defence of our own territory and [has] no aggressive intentions whatsoever."[130]

The second approach was to open up a dialogue with the Chinese government. This tactic was outlined in an internal India Office minute: "We need not perhaps speak very sharply the first time round, and indeed we need to [do] little more than express our hope that the rumours of Chinese military movements are untrue and ask for the confirmation of the Chinese that they are without foundation."[131] In response, Wu Guozhen, the Acting Foreign Minis-

---

**123** TBL, IOR/L/P&S/12/2175, telegram from the Indian government to India Office, July 27, 1943.

**124** TBL, IOR/L/P&S/12/4210, minute by H.A.F. Rumbold, April 22, 1943.

**125** Lamb, *Tibet, China & India, 1914–1950*, 317.

**126** TBL, IOR/L/P&S/12/4210, telegram from Ludlow (Gangtok) to Gould (New Delhi), April 8, 1943.

**127** Telegram from *Chargé d'affaires* in India to the Secretary of State, May 15, 1943. US Department of State, ed., *FRUS: 1943 China*, 631.

**128** Telegram from *Chargé d'affaires* in India (Merrell) to the Secretary of State, May 14, 1943, ibid., 629–630.

**129** TNA, FO 371/35755, telegram from External Affairs Department (GOI) to India Office, May 17, 1943.

**130** TNA, CAB 111/336, telegram from External Affairs Department (GOI) to India Office, June 7, 1943.

**131** TBL, IOR/L/P&S/12/4210, minute by H.A.F. Rumbold, April 22, 1943.

ter, reassured the British that there was no Chinese movement against Tibet, but he also emphasised that "a foreign government had no more right to interest themselves in Tibetan affairs than in those of Provinces of China proper."[132] Meanwhile, Chongqing used private channels to indicate to the Kashag that military intervention was still a possibility in order to keep up the pressure.[133]

The third approach was to seek to obtain American support. The increasing Anglo-Chinese tensions over Tibet were evident in Washington in March 1943 and at a high-level meeting of the Pacific War Council (PWC) in May, they came to the surface. Song Ziwen complained to Anthony Eden about the attitude of the British-Indian government to China's concerns over the political status of Tibet and the supply routes. Since Eden was not very familiar with the Tibetan situation, the discussions did not go very far.[134] Yet as the rumours of a Chinese invasion intensified, the dispute came to involve Churchill, Roosevelt and Song (on behalf of Chiang Kai-shek). At a meeting of the PWC on May 20, with Roosevelt and Song in attendance, Churchill denounced China as in "danger of ... becoming involved in imperialistic adventures."[135] The following exchanges are recorded in a Chinese source:

> Prime Minister: I hear Chinese troops are mobilized for action against Tibet and this provoked in that independent country a very great alarm. I would like to have assurances from the Chinese Government that no untoward action will be taken.

> Dr Soong: I know of mobilization of Chinese troops against Tibet and I don't think there is any case for alarm. But the Prime Minister referred to Tibet as an 'independent country.' May I remind the Prime Minister that China has many treaties with the British Empire in regard to Tibet which is acknowledged as coming under Chinese sovereignty.

> Lord Halifax[136]: Is it not rather Chinese 'suzerainty'?

> Dr Soong: It is part of the Chinese Republic. Tibet is a religious hierarchy and not an independent country, as recognised in Sino-British treaties.

---

132 TBL, IOR/L/P&S/12/4210, telegram from Chongqing to Foreign Office, May 8, 1943.
133 May 12, 1943, see Lin, ed., *The Diary of Dr. Wang Shih-chieh*, 506.
134 TNA, PREM 3/90/5B, message from Eden to Churchill, enclosed in the telegram from Washington to Foreign Office, March 16, 1943; TBL, IOR/L/P&S/12/4194, telegram from Viscount Halifax to Foreign Office, March 18, 1943.
135 TNA, PREM 3/90/5B, letter from Prime Minister to Foreign Secretary, May 22, 1943.
136 Lord Halifax was then the British Ambassador to the United States (1940–1946).

> The Prime Minister: We ourselves are not interested in Tibet; it is a bleak, forbidding coun
> try. We are only anxious that energies be not dissipated; they should be concen
> trated against the common enemy.[137]

Song then enquired about Roosevelt's views on the status of Tibet. Roosevelt apparently replied that

> I asked Churchill what on earth he meant and if he had any interest in Tibet. He replied he
> had none at all. Then I asked him: why should he interfere in the Chinese affairs. Churchill
> replied that certainly he had no intention of interfering, as I told you once before[.] I used
> Jimmy's expression and asked him 'and so What?' Churchill is thinking of having Tibet [as]
> a barrier against both Russia and China.[138]

The PWC conference ended without any firm backing from the Americans, but
the British were confident that any diversion of Chinese military resources to
an "adventure in Tibet" would be objected to by the Americans.[139] However,
with regard to the political status of Tibet, there were doubts as to whether
the Americans were in concert with the British. In an aide-memoire to the British
Embassy at Washington on May 15, 1943, the American stance was made evident:

> With regard to the position of Tibet in Asia, ... the Government of the United States has
> borne in mind the fact that the Chinese Government has long claimed suzerainty over
> Tibet and that the Chinese constitution lists Tibet among areas constituting the territory
> of the Republic of China. This Government has at no time raised a question regarding either
> of these claims.[140]

# 4 American Concerns over Tibet

The importance of the trans-Tibetan supply route to the Allied war effort persuaded the Americans to take a closer interest in Tibetan affairs. The newly-
formed Office of Strategic Services (OSS), a predecessor of the Central Intelligence Agency (CIA), took an important role here. Two military officials, Lieutenant Brooke Dolan II and Captain (Count) Ilya Tolstoy, a grandson of the great

---

**137** Meeting of the Pacific War Council, May 20, 1943. These dialogues were recorded by a Chinese secretary in English. Quoted from Wu and Kuo, eds., *T. V. Soong: Selected Minutes of Meetings with Foreign Leaders, 1940–1949*, 341.
**138** Conversation between the President and Dr. Soong, June 4, 1943. See ibid., 362–363.
**139** TBL, IOR/L/P&S/12/4210, telegram from Chongqing to Foreign Office, May 8, 1943.
**140** Aide-memoire from the Department of State to the British Embassy, May 15, 1943. US Department of State, ed., *FRUS: 1943 China*, 630–631.

Russian novelist Lev Tolstoy, were sent to Tibet by the OSS in early 1943. The main American interest at this time was the possibility of establishing airfields and the feasibility of building a road to connect Jyekundo (Yushu 玉樹) in Qinghai with Lhasa.[141] However, more interesting was the sympathetic attitude towards Tibet's independence.

One inadvertent consequence of the Anti-Japanese War was the westward advance of Chinese Nationalist influence. The relocation of the Chinese capital from Nanjing to Chongqing shortened the geographical distance between the Nationalist-dominated areas and Tibet. This made it possible for the Guomindang, for the first time since the Qing government collapsed in 1912, to wield strong influence over Tibetan issues. The Kashag was worried about this. Ringang told Ludlow in February 1943 that

> the Tibetan government was very uneasy at the present attitude of the Chinese government, and that he feared trouble as soon as the war was over. He said some Tibetans were so afraid of what China would do in the near future that they were almost tempted to wish for a Japanese victory over China, though not an Axis victory over the Allies.[142]

He further remarked that what the Tibetan government hoped for was an Anglo-American assurance over the independence and integrity of Tibet. Clearly, Captain Tolstoy was influenced by these Tibetan anxieties. He recommended to his government that Tibet should be separately represented at a future peace conference. This was welcomed by the Tibetans and by Ludlow.[143]

Tolstoy's sympathies encouraged the Tibetans but troubled the British-Indian authorities. At the time, the British had no intention to internationalise the matter of Tibet. The Governor-General and Viceroy of India, the Marquess of Linlithgow, deemed the American efforts "amateur". Linlithgow was also disappointed at Ludlow's endorsement of this proposal. Nor was he impressed by suggestions from others such as Basil Gould, the Political Officer in Sikkim, and Hugh Richardson, the head of the British Mission to Lhasa, that Tibet should be prompted to follow up contact with the Americans and send a letter to Washington outlining their views on independence.[144] It transpired that the British desire was not to endorse Tibet's independence but simply to call a halt to Chinese

---

141 Lamb, *Tibet, China & India, 1914–1950*, 315.
142 TNA, FO371/35755, copy of a demi-official letter from F. Ludlow to Sir Basil Gould, February 20, 1943.
143 TNA, FO371/35755, copy of a demi-official letter from F. Ludlow to Sir Basil Gould, March 14, 1943.
144 TBL, IOR/L/P&S/12/4210, telegram from Viceroy to India Office, May 3, 1943.

aggression. In the end, American interest in Tibet dwindled. By December 1943, 10,000 tons of goods per month were being transported to China and as the capacity of the airlift over "the Hump" increased, so the US lost interest in Tibet.[145] Only with the Korean War in the 1950s did Tibet once again attract American attentions, as a leverage to confront Chinese Communism.[146]

## 5 The British Review Their Tibetan Policy in 1943: the Concept of a Buffer

Sino-British negotiations over Tibet resumed in July-August 1943, as Song Ziwen and Anthony Eden had agreed in Washington to continue talks when Song visited London. Thus, it was necessary for the British government to re-examine its Tibetan policy before that date. Discussions between London and New Delhi therefore commenced.

British interests in Tibet were mainly attributed to security concerns over India. It was conceived as an important buffer against China and Russia, the value of which had been outlined by Charles Bell in the 1920s.[147] Bell, a former head of the British Political Office in Sikkim and a Tibetan specialist, argued that

> We want Tibet as a buffer to India on the north. Now there are buffers and buffers; and some of them are of very little use. But Tibet is ideal in this respect. With the large desolate area of the Northern Plains controlled by the Lhasa government, central and southern Tibet governed by the same authority, and the Himalayan border states guided by, or in close alliance with, the British-Indian government, Tibet forms a barrier equal, or superior, to anything that the world can show elsewhere.[148]

In 1940, Olaf K. Caroe, the Foreign Secretary of the British-Indian government, had evolved this into a theory of two buffer belts. The inner buffer belt consisted of, from west to east, the Baluch and Pathan tribal areas, Nepal, Sikkim, Bhutan,

---

**145** John D. Plating, *The Hump: America's Strategy for Keeping China in World War II* (College Station: Texas A&M University Press, 2011), 164.

**146** Qiang Zhai, "Tibet and Chinese-British-American Relations in the Early 1950s," *Journal of Cold War Studies* 8, no. 3 (2006), 44–52.

**147** Charles Bell was the Political Officer in Sikkim (May-December 1904, 1906–07, 1908–11, 1911–13, 1914–18 and 1920–1921).

**148** Sir Charles Bell, *Tibet: Past and Present* (Oxford: The Clarendon Press, 1924), 246. The British officials frequently quoted Bell's words. See TBL, IOR/L/P&S/12/4195A, telegram from the Secretary of the External Affairs Department (GOI) to India Office, September 19, 1945.

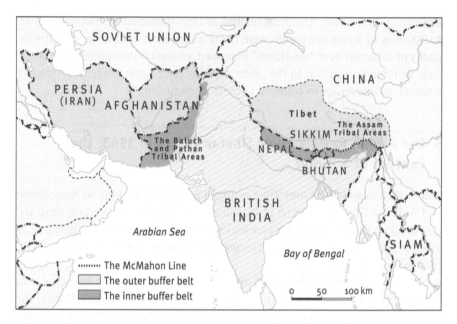

**Map 8:** Two Buffer Belts.
© Zhaodong Wang & © Peter Palm, Berlin.

and the Assam tribal areas. An outer buffer belt included Persia, Afghanistan and Tibet (see map 8).

The inner and outer belts were partly connected. Nepal, Bhutan, Sikkim and the Assam tribal areas, for example, had close racial, religious and commercial links with Tibet and were thus described by Caroe as "the Mongolian Fringe".[149] In his view, it was imperative to dominate these belts in order to prevent any possible occupation of India by hostile powers. Otherwise, significant military resources would have to be spent on securing the 1,000-mile frontier between India and Tibet. If, for example, Tibet was under direct Chinese control, as the India Office speculated, it could become an uncomfortable neighbour to India. The Chinese government might choose to present Indo-Tibetan territorial disputes as being "contingent on concessions in a wider diplomatic sphere." Moreover, the Guomindang might pick up the former Qing court's claims to suzerainty over Nepal, which was a vital recruiting ground for the Indian Army. This sce-

---

149 TBL, IOR/L/P&S/12/4194, "the Mongolian Fringe", note by Foreign Secretary of GOI (O.K. Caroe), January 18, 1940.

nario could conceivably be extended to Sikkim and Bhutan as well.[150] Therefore, in Caroe's words, the main British objective should be "to take advantage of [British] geographical position to keep the window [of Tibet] open" and reinforce the line "in order to prevent Lhasa falling into enemy hands."[151]

## 5.1 Discussions over Chinese Suzerainty

Chinese attempts in 1942–1943 to exploit the trans-Tibet supply lines in favour of their sovereignty claims to Tibet, despite being successfully blocked by the British, convinced London of the fact that the Nationalist annexation of Tibet was inevitable once China restored peace and united. In this case, it was necessary to review Britain's Tibet policy and decide whether Britain should continue or refrain from the support of Tibet's independence.

British diplomats specialising in Chinese affairs, exemplified by Eric Teichman, argued that in the future, "Tibet proper will no doubt ultimately return to the Chinese fold as a self-governing dominion of the Chinese Commonwealth"; and it would "be wise to promote rather than obstruct the development of Tibet along these lines." He also noted that there were no further Chinese ambitions with regard to Nepal, Sikkim, or Bhutan.[152] However, the India Office and the British-Indian government demurred. R.T. Peel, an official in the India Office, insisted that Tibet's status as a buffer must be retained as it served India's interests eventually:

> If China should become a united and aggressive power Tibet's importance as a buffer for India would increase, although so would the difficulty of preventing her absorption by China. If China should become 'rent' with civil war once again, it would be desirable that Tibet should remain untouched by these disturbances, while if China made peace with Japan and fell under Japanese control, we would wish to prevent that control extending to Tibet.[153]

Thus, he suggested that London should take measures to prevent Tibet from being annexed by China. The first, and the most important, was to abandon the British recognition of Chinese suzerainty over Tibet, which had been manifested in a formal declaration to the Chinese government in 1921. The commit-

---

150 TBL, IOR/L/P&S/12/4194, memorandum by unknown person, unknown date.
151 TBL, IOR/L/P&S/12/4194, "the Mongolian Fringe", note by Foreign Secretary of GOI (O.K. Caroe), January 18, 1940.
152 TBL, IOR/L/P&S/12/4194, letter from E. Teichman to H. Weightman, July 27, 1942.
153 TNA, FO 371/35755, letter from R.T. Peel to Ashley Clarke, May 8, 1943.

ment, in the views of the British India experts, was originally introduced to distinguish Tibet from China but now gave the Chinese some grounds to claim control over Tibet.[154] Furthermore, it "might prevent [the British] at the peace settlement from supporting a Tibetan claim to full independence."[155]

For its part, London had broader considerations. It had been agreed by the Foreign Office and the India Office at a joint meeting in May 1943 that the present time was inopportune to give Tibet effective material support when Britain was allied with China in a major war.[156] Furthermore, the American attitude was unclear and British support for Tibet could not continue if the United States sided with China.[157] Moreover, a sudden reversal of British Tibetan policy might infuriate China and precipitate the latter's incorporation of Tibet. Based on these factors, Ashley Clarke, the head of the Far Eastern Department of the Foreign Office, suggested that a better course of action might be to "insist on Tibetan autonomy as being a condition of our recognition of Chinese suzerainty."[158]

Two dilemmas were implicit in this approach: whether an explicit definition should be given to the "suzerainty", and whether the British should agree to an exchange of legations between India and Tibet in light of the establishment of Tibet's foreign affairs department. In regard to the first issue, "suzerainty" had been an ambiguous term since it had been introduced in the Tibetan case by George Curzon, the Viceroy of India, in the early 1900s as a tool to differentiate Tibet from China. During the ensuing half a century, the term had meant little in practical terms. As R. T. Peel put it:

> At the time of our invasion of Tibet in 1904, we treated [Tibet] as virtually independent of China, while both before and after this incident we attempted to deal with questions affecting Indo-Tibetan relations by direct negotiations with the Chinese government and without reference to the Tibetans. Since 1912 we have been consistent in treating China's suzerainty as little more than an empty form. ... Since the breakdown of negotiations in 1914 we have, for all practical purposes, dealt with Tibet as if it was an international unit quite separate from China. We have, for example, sold her arms and ammunition, maintained a representative at her capital, acted as an intermediary on occasion between Tibet and her suzerain, and negotiated with the Tibetan government the 1914 Trade Regulations without reference to China. When the Memorandum of 26th August 1921 ... was handed to the Chinese Minister in London, an oral communication was made to him at the same time to the effect that we regarded ourselves as at liberty to deal with Tibet as an autonomous state, if necessary

---

154 TBL, IOR/L/P&S/12/4194, minute by H.A.F. Rumbold, May 22, 1943.
155 TNA, FO 371/35755, telegram from Ashley Clarke to Horace Seymour, July 29, 1943.
156 TNA, FO 371/35755, minute by Ashley Clarke, May 15, 1943; TBL, IOR/L/P&S/12/4194, points for discussion at meeting on May 18, in India Office, not dated.
157 TBL, IOR/L/P&S/12/4194, a draft letter from R. Peel to Caroe, June 11, 1943.
158 TNA, FO 371/35755, telegram from Ashley Clarke to Horace Seymour, July 29, 1943.

without further reference to China, to enter into closer relations with her, to send an officer to Lhasa from time to time and to give the Tibetans any reasonable assistance they might require in the development and protection of their country.[159]

The academic meaning of suzerainty was nominal sovereignty over a semi-independent or internally autonomous state and might be presumed to be interpreted as precluding Tibetan rights to maintain separate foreign relations of any kind.[160] Yet from the India Office's perspective, this definition did not apply to British dealings with Lhasa through the newly established TFB. Thus an "elastic" suzerainty without clear restriction seemed to be a wise option.[161] As for the second issue, Peel deemed a legation unnecessary. In his opinion, it might be embarrassing to host a Tibetan agency in India if the Chinese conquered Tibet or the Tibetans reached a settlement with China as part of which they recognised Chinese suzerainty.[162]

In the end, an agreement on the way forward was achieved in June 1943 and this was ratified by the War Cabinet. The main points were as follows:

> We should state categorically that neither His Majesty's Government nor the Government of India has any ambitions in Tibet other than the maintenance of friendly relations.
>
> We should recall that our attitude has always been that we recognise Chinese suzerainty, but that this is on the understanding that Tibet is regarded as autonomous. We should avoid any unconditional admission of Chinese suzerainty. ...
>
> If the Chinese Government contemplate the withdrawal of Tibetan autonomy, His Majesty's Government and the Government of India must ask themselves whether in the changed circumstance of today it would be right for them to continue to recognise even a theoretical status of subservience for a people who desire to be free and have, in fact, maintained their freedom for more than 30 years.[163]

## 5.2 The Song-Eden Meetings in London, July-August 1943

During his visit to London in the summer of 1943, Song Ziwen held discussions with Anthony Eden on three occasions and with Churchill once. His disagreements with Eden over Tibet revolved around the definition of "suzerainty". He pointed out that "Great Britain had always recognized Tibet as a part of Chinese

---

**159** TNA, FO 371/35755, letter from R.T. Peel to Ashley Clarke, May 8, 1943.

**160** TBL, IOR/L/P&S/12/4210, telegram from External Affairs Department (GOI) to India Office, June 1, 1943.

**161** Ibid.

**162** TNA, FO 371/35755, letter from R.T. Peel to Ashley Clarke, May 8, 1943.

**163** TNA, CAB 66/38/17, "Status of Tibet" memorandum by FO and IO, June 23, 1943.

territories and also admitted Chinese suzerainty over it." Gu Weijun, the Chinese ambassador, reinforced this:

> autonomy did not always exist in Tibet. At one time the administration there was under the supreme control of the Chinese Government. The degree of the control of the Chinese Government over Tibet fluctuated from time to time. China considered that autonomy was [an] internal question between her and the Tibetan authorities, and that any outside influence would merely complicate the relations between Tibet and the Chinese authorities.[164]

But Eden maintained that "Great Britain held a different view about the suzerainty of China over Tibet."[165] Song chose not to pursue the matter, possibly because his main aim for this trip was to expedite the start of the Burma counterattack and he thus had no wish to let the question of Tibet dominate the discussion.[166] Nevertheless, he expressed the hope to Eden that there would be no difficulty when the time came for a settlement. Meanwhile, he remarked that "he could not conceive that Tibet was of great importance to Great Britain, whereas to China it was part of her territory."[167]

On his return to Washington, Song raised the matter again. In a meeting with both British and American officials in September 1943, he remarked that Tibet was an integral part of China and the question of Sino-Tibetan relations was a purely Chinese domestic matter. He suggested that the British should not regard Tibet as an area or a people independent of China. He acknowledged that by virtue of geography, British India did have a special interest in Tibet but affirmed that politically, and legally, Chinese claims stood on much firmer ground than did those of the British.[168]

---

**164** Resume of a conversation with Mr. Eden, British Foreign Secretary at Foreign Office, July 26, 1943, Wu and Kuo, eds., *T. V. Soong: Selected Minutes of Meetings with Foreign Leaders, 1940–1949*, 375.
**165** Ibid.
**166** Koo, *Gu Weijue huiyilu vol. 5*, 327–328.
**167** Notes of a Conversation between Dr. Soong and Dr. Koo on one hand, and Mr. Eden on the other, August 11, 1943, Wu and Kuo, eds., *T. V. Soong: Selected Minutes of Meetings with Foreign Leaders, 1940–1949*, 409–411.
**168** The Acting Secretary of State to the Ambassador in China (Gauss), September 29, 1943, US Department of State, ed., *FRUS: 1943 China*, 641–642.

## 6 A Further British Dilemma: Practical Control of the Assam Border Areas

Apart from the link between Chinese suzerainty and Tibetan autonomy, another component for British Tibetan policy in 1943 was to expedite practical control of the Assam Himalayas or South Tibet (Zangnan 藏南). This operation aimed to prevent possible Chinese moves to claim these territories.

The Sino-Indian, or Indo-Tibetan, disputes over these territories can be traced back to the Simla Conference. Before then, the traditional Indo-Tibetan boundary, or the Outer Line, stretched from the southeast point of Bhutanese border along the edge of the Himalayan foothills across the Lohit River to meet Burma in the Hkamtilong region. This traditional line was designated under an Anglo-Tibetan accord of 1872–73 (see map 9).[169] One aim of the

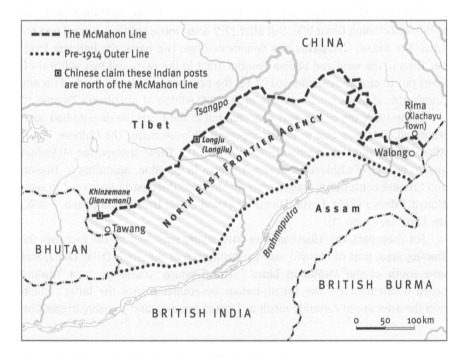

**Map 9:** The Sino-Indian Boundary Dispute (Eastern Sector).
Based on Lamb, "The Present Boundary Dispute Eastern Sector," in Lamb, *The China-India Border*, map 2, 10. © Zhaodong Wang & © Peter Palm, Berlin.

---

169 The line can be found in the map appended to Vol. II of the 1909 edition of Aitchison's *Treaties*. See ibid., *Tibet, China & India, 1914–1950*, 401–402.

Simla Conference had been to consolidate this boundary to India's advantage. Indeed, Henry McMahon contemplated a new border, moving the traditional line northwards by roughly 100 miles along the line of the Himalayan mountains. This was named the Red Line or the McMahon Line (see map 9). As we have seen, McMahon induced the Tibetans to accept this border by promising diplomatic mediation over the Sino-Tibetan border. The new border was confirmed by two exchanges of notes between McMahon and Lönchen Shatra in March 1914, during the Simla Conference and without reference to China. Tibet's agreement was contingent on its Greater Tibet plan being fulfilled by the British. This involved incorporating Chinese-controlled Amdo and Kham into the state territory of Tibet. However, Lönchen Shatra, an inexperienced diplomat, did not put this precondition into the exchanges of notes. For their part, the Chinese had very good reason to disregard the new border, since they had neither signed nor authorised the Tibetan government to sign these treaties. After the Simla Conference, little attention was paid by the British to the McMahon area due to the forthcoming Great War. But after 1919 some loose political control was exerted. The Assam Himalayas was demarcated into two parts, the Balipara Frontier Tract in the west and Sadiya Frontier Tract in the east, and two political officers put in charge (see map 10).[170] Yet the new border was not marked on any British official maps, such as those in the 1929 edition of Aitchison's *Treaties*.[171]

In the 1920s the border was peaceful, but in the 1930s the dispute had been rekindled. This was due to newly approved Chinese maps: the Chinese Postal Atlas of 1933 and the *Shenbao* 申報 Atlas of 1934. These showed the McMahon area to be part of Chinese territories. The area in question, according to Tibetan and Chinese convention, was roughly divided into three parts from west to east: Monyül (Menyu 門隅), Loyül (Luoyu 珞隅) and Lower Dzayül (Xia Chayu 下察隅, the Lhot) (see map 11).

For their part, the Tibetans were particularly exercised by the fact that the Tawang area, part of Monyül and the birthplace of the Sixth Dalai Lama, was now south of the McMahon Line.[172] The Tibetans' insistence over Tawang posed a dilemma for the British-Indian government since the latter's claim over the area would certainly result in Tibetan protests, and possibly trigger Chi-

---

170 Ibid., 409–410.
171 Olaf K. Caroe spotted this and managed to revise it in 1938 by publishing a new 1929 version to substitute the old one. See ibid., 415.
172 Robert Reid, *History of Frontier Areas Bordering on Assam 1883–1941* (Delhi: Eastern Publishing House, 1983), 287–288.

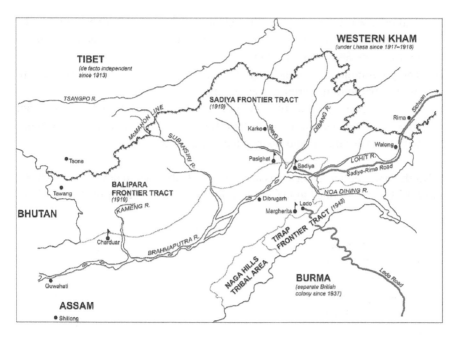

**Map 10:** The Eastern Himalayas, c. 1943.
Made by Bérénice Guyot-Réchard and Tina Bone, and taken, with permission, from Guyot--Réchard, *Shadow States*, map 5, 278.

nese involvement, while if the British did not stand their ground the validity of the entire 1914 convention might be open to question.[173]

In 1943 the issue of the Indo-Tibetan border was revisited as part of the discussions over Chinese suzerainty. Given the possibility that China might incorporate Tibet into its territories after the war, it became urgent for the British to control these areas. In April 1943, the matter was discussed in detail by the Foreign Office and the India Office. The former was worried that British penetration into these areas, which had hitherto been left alone, would produce Chinese protest; and that China would probably receive American support. It was thus concluded that extreme caution should be exercised in formulating any active policy towards the Indo-Tibetan borderland.[174] The War Cabinet approved a plan along the following lines:

---

173 Lamb, *Tibet, China & India, 1914–1950*, 418.
174 Ibid., 457.

a. Authority was given to gradually and cautiously extend British control to the McMahon Line. No British action should be taken for the time being to the north of the Sela Pass [the Tawang area].
b. The programme of "vindicating" the McMahon Line should be getting underway without any prior consultation with the Tibetans.
c. It was emphasised that the new policy in the Assam Himalayas must not soak up troops who were needed for the war against Japan; armed clashes with the Tibetans should be avoided at all costs.[175]

A North-East Frontier Agency was accordingly established to devise and then implement this new policy along the frontier areas to the north of the old Outer

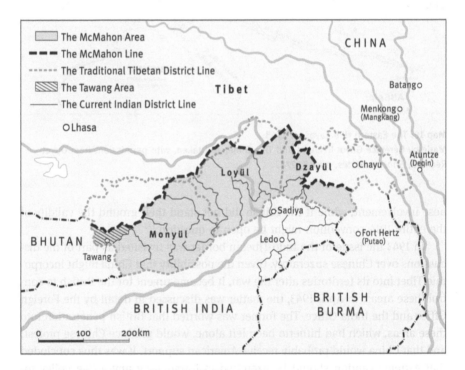

**Map 11:** Dzayül, Loyül, Monyül and the McMahon Area.
According to the present administrative division of the PRC, the three areas are approximately equivalent to three counties from west to east: Cuona county (錯那縣), Motuo county (墨脫縣) and Chayu county (察隅縣). The Motuo and Chayu belong to the city of Linzhi (林芝市) while Cuona is part of the city of Shannan (山南市). But the territories south of the McMahon Line are actually controlled by India. © Zhaodong Wang & © Peter Palm, Berlin.

---

175 Ibid., 458–459.

Line. In Lower Dzayül, for example, the villages of Walong (Wanong 瓦弄, on the right bank of the Lohit) and Tinai (on the left bank) were populated by Tibetans but several military posts were created there. The Foreign Office was extremely anxious about any armed conflict being aroused by these incursions. An armed incident between the Indians and Tibetans would "inflame the Chinese suspicions" and possibly line up the Tibetans, the Chinese, and the Americans against the British.[176] Yet, fortunately for the British, there was little opposition. In Loyül, the British created military posts at Karko and Riga. In Monyül, the garrison at Rupa was augmented, and a new post was established at Dirangdzong (Derangzong 德讓宗). The land north of the Sela Pass was left strictly alone and nowhere in the Tawang area was made a formal declaration of British sovereignty.[177]

In parallel with these infiltrations, the British put forward proposals to the Tibetans when Basil Gould visited Lhasa in 1944. The first was that the British abandon Tawang in exchange for Tibetan commitments regarding other parts of the McMahon area. The second was Tibet's concession of the whole of the McMahon area to Britain in return for financial compensation to the Drepung monastery (Zhebangsi 哲蚌寺) in Lhasa for the loss of revenue to its Monyül counterpart as a result of British actions in that area. However, the Kashag accepted neither of these proposals.[178] The Tsongdu summed up the mood:

> We look upon the British Government with confidence and for assistance, but the British Government have occupied indisputable Tibetan territory by posting British officers and troops and have said that they could not withdraw them from these areas. The Sino-Tibetan question which is being negotiated with the British Government as an intermediary, has not yet been settled, and the areas mentioned above have not been shown in the treaty as included within Indian territory. This question has never been raised since the Wood Tiger Year (1914) which is now 30 years ago. Considering this fact, we regret to say that we cannot agree with the Government of India's action in taking these Tibetan areas within British territory. If the officers and troops that are posted at Kalaktang and Walong are not withdrawn immediately, it will appear like a big insect eating up a small one, and the bad name of the British Government will spread like the wind.[179]

---

**176** TBL, IOR/L/P&S/12/4214, letter from J.C. Sterndale Bennett (FO) to F.A.K. Harrison (IO), October 6, 1944.
**177** Lamb, *Tibet, China & India, 1914–1950*, 464–465.
**178** TBL, IOR/L/P&S/12/4194, points arising from discussion with Tibetan Foreign Office, September 16, 1944.
**179** TBL, IOR/L/P&S/12/4223, telegram from Sir Basil Gould to Indian Government, June 19, 1945. Quoted in Lamb, *Tibet, China & India, 1914–1950*, 468.

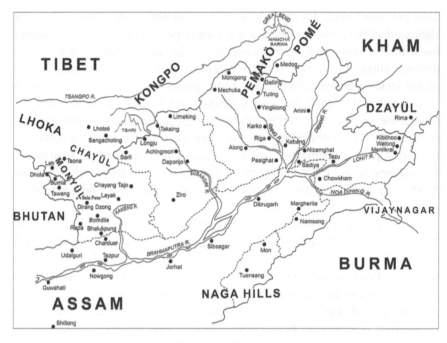

**Map 12:** Main Places in the McMahon Area.
Made by Bérénice Guyot-Réchard and Tina Bone, and taken, with permission, from Guyot-Ré-chard, *Shadow States*, map 2, 275. A small error about the location of 'Sela Pass' in the original map has been rectified with the permission of the copyright holders.

Despite these objections, a skeleton-like frontier administration was developed in the Assam Himalayas as the war ended. This included some permanent outposts, a number of airfields (six of eighteen planned airfields were put in practice) and a road from Sadiya to Walong.[180] During this process, Paul Mainprice, the Political Official for the Lower Dzayül (the Lohit Valley), even suggested pushing the McMahon Line further north to encompass Upper Dzayül , where rice-growing lands, a sizeable population and abundance of flat land were located.[181] But at the end of the war, the British-Indian government was preoccupied with a more urgent matter: the question of independence and antagonism between Hindus and Muslims in India. Nonetheless, the wartime developments paved the way for the Assam Himalayas (the Assam Frontier) to be incorporated

---

**180** New Delhi, NAI, External Affairs Proceedings (1946), Post-war maintenance of airfields on the NEF, 61-NEF/46, quoted in Guyot-Réchard, *Shadow States*, 78.
**181** New Delhi, NAI, External Affairs Proceedings (1946), Ratification of the McMahon Line below Rima, 307-CA/46 (Mainprice to Mills, September 12, 1945), quoted in ibid., 77.

into an independent Union of India. Accordingly, in the Indian Constituent Assembly a "North-East Frontier (Assam) Tribal and Excluded Areas Sub-Committee" was created to represent the Assam Himalayas.[182] In November 1948, a "Sixth Schedule" was adopted to give Assam's hill tribes a series of political and socio-economic safeguards, including the formation of Autonomous Districts and Regions headed by councils with a wide range of powers.[183]

## Conclusion

The Second World War placed the British in a tricky position because, on the one hand, they believed the trans-Tibet supplies were vital to the war effort but, on the other hand, they worried such linkage would assist Chinese rights regarding Tibet that Britain had long opposed. This position was due to contradictory opinions held by two branches of the British government respectively. One was the Foreign Office and the British Embassy to China, and the other was the India Office and the British-Indian government. Britain's conflicts with China over Tibet unexpectedly drew the attention of the United States to the matter and forced the latter to declare their hand, which to some degree confirmed China's sovereignty over Tibet.

The Nationalists were intransigent over their claimed sovereignty rights over Tibet. It was evident that Chongqing attempted to advance its sovereignty claims over Tibet by exploiting the Allied concerns of opening trans-Tibet supply lines. Although these efforts were thwarted by the Tibetans with the British support, London was convinced that the incorporation of Tibet into Chinese territory was inevitable once the war ended and China was united. If it was the case, the intractability of the ambiguity of "suzerainty" and the territorial disputes with Tibet must be dealt with in advance. However, the Pacific War prevented London from any abrupt policy reversal. Consequently, the British preserved the recognition of Chinese suzerainty over Tibet despite tying it to a vague Chinese promise of Tibet's autonomy. In fact, it acknowledged the Chinese having the right to assume some degree of control over central Tibet where little Chinese authority existed in practice at the time. At the same time, British ambitions contracted to focus on the disputed McMahon area (the Assam Himalayas). It was believed that Chinese territorial claims would certainly extend to these territories

---

[182] In the Government of India Act (1935), these territories were partially or totally excluded from Indian territories.
[183] Guyot-Réchard, *Shadow States*, 77–83.

after resuming power in Tibet and, thus, the British government was advised to quickly and practically occupy them regardless of the war. These territories were later inherited by independent India and considerably assisted the latter's claim to these territories.

For its part, Lhasa had no intention to be involved in the war but yielded to either the Chinese or the British pressures. The war not only strained Tibet's relations with China but also with Britain. In the context, some British Tibet specialists were convinced that only stronger British support for Tibetan independence could ameliorate the deteriorated Anglo-Tibetan relations and deter Tibet from coming to terms with China. A lobbying war thereafter commenced between the Chinese and British governments over the raising or dampening down of a Tibetan spirit of independence. This will be discussed in the next chapter.

# Chapter VI
# The Last Sino-British Rivalry for Tibet, 1944–1947

The Sino-British contest over Tibet continued into the post-war era. The context was, however, changing: the war was approaching its end and Tibet lost its strategic role as a passage to China; British undertakings in the Assam Himalayas strained its relations with Tibet and drove the latter to seek either an accommodation with China or international support; and India was to become an independent nation whose security was no longer a British concern. Against this backdrop, Chongqing (Nanjing after May 5, 1946) and London developed their respective post-war Tibetan policies. Some academic works have examined these policies but given relatively little heed to China-Britain rivalries over Tibet during this period.[1] This chapter will seek to bridge the gap. It will explore China's efforts to counter British initiatives in Tibet, its efforts to absorb Tibet into its post-war constitutional framework and the responses of the Foreign Office and the India Office to these challenges. The year 1947 was a watershed for both Britain's and China's Tibetan strategies. In that year the Nationalist intelligence network in Tibet was exposed and the regime lost much influence among pro-Chinese Tibetans, while a newly independent Indian government inherited British treaty rights in Tibet. Meanwhile, the United States began to reformulate its approach to Tibet as the Cold War began to heat up.

## 1 Anglo-Chinese Manoeuvring over Tibet, 1944–1945

As we have seen, Britain's wartime Tibetan policy created apprehension in China. At the Tibetan Affairs Conference of the Nationalist government in December 1944, Liang Long 梁龍, the head of the European Department in the Chinese Foreign Ministry, listed eight instances of British aggression in Tibet:

1. British supplies of weapons and ammunition to Tibet;
2. British advance on Tawang;

---

[1] See Lamb, *Tibet, China & India, 1914–1950; ibid., McMahon Line (I); ibid., McMahon Line (II); ibid., The China-India Border; Lin, Tibet and Nationalist China's Frontier; Goldstein, Demise of the Lamaist State; Smith Jr., Tibetan Nation; Chen, Kangzhanqianhou zhi Zhong-Ying Xizang jiaoshe; Zhu, Xizang zhuanshi.*

https://doi.org/10.1515/9783110706659-013

3. British aggressions in the Lower Dzayül;
4. British sales of wireless radio sets to Tibet;
5. British road construction in the Assam Himalayas[2];
6. British exploration of oil fields in Tibet;
7. British propaganda encouraging Tibetan independence;
8. An English school established by the British near Lhasa.[3]

These intrusions galvanised the Chinese into action. A new round of manoeu-vrings thus began between Britain and China.

## 1.1 Propaganda and Munitions

In British eyes, Tibet had not entirely separated itself from China. As one British official observed, while Tibetan official pronouncements made it clear that "the Tibetan Government had not admitted Chinese suzerainty", felicitations had still been conveyed to Chiang Kai-shek on his appointment as President of the Chinese Republic in which Chiang was described as "Our President" and Tibet was referred to as part of China.[4] Thus, Indo-Tibetan disputes over the McMahon area carried a risk of pushing Tibet into the Chinese camp. To avoid this, British officials aimed to strengthen the link between India and Tibet through such initiatives as the creation of an English school in Lhasa[5]; by broadcasting in the Ti-

---

2 Specifically, the Chinese were convinced that at least three roads were under construction: (a) the road from Gangtok, the Capital of Sikkim, to Ledula 勒都拉, a border town of Tibet; (b) the road from Trashigang in Bhutan to Tawang; (c) the road from Sadiya to Walong. See respectively IMH Archives, 11–29–26–00–024, *Guanyu Zang'an huiyi* [The Conference concerning Tibet], minutes of The Conference concerning Tibet, December 11, 1944; IMH Archives, 11–29–26–00–015, *Ying dui Zang xuanchuan duli* [British Propaganda to Encourage Tibetans for Independence], telegram from MAC to Foreign Ministry, November 1, 1945; IMH Archives, 11–32–11–02–010, *Yingyin qinlue Xikang bianjing* [The British-Indian Invasion to the Border of Xikang], telegram from MAC to Foreign Ministry, October 9, 1944.
3 IMH Archives, 11–29–26–00–024, *Guanyu Zang'an huiyi* [The Conference concerning Tibet], minutes of The Conference concerning Tibet, December 11, 1944. The location of the school in this Chinese document seems to be wrongly recorded as Dekyi Lingka (Dejilinga 德基林噶 or De-jilinka 德吉林卡). The planned location was Lubu 魯布 near Lhasa.
4 TBL, IOR/L/P&S/12/4194, extract from memorandum dated July 9, 1944.
5 The creation of an English school, however, did not progress well due to the opposition of the three largest monasteries at Lhasa, the Sera (Sela 色拉), Drepung and Ganden (Gandan 甘丹): very few students registered. See IMH Archives, 11–29–26–00–015, *Ying dui Zang xuanchuan duli* [British Propaganda to Encourage Tibetans for Independence], telegram from Foreign Ministry to MTAC, December 27, 1944. The English school referred to here was established in 1944 in

betan language from Gangtok in Sikkim; and by modernising communications with Tibet via Sikkim.[6] The British also had plans to set up a British hospital in Lhasa to rival its Chinese counterpart.[7]

In addition, 250 newspapers were distributed every month by the British-Indian government from Kalimpong to Tibetan nobles at Yatung, Shigatse and Lhasa, and these highlighted British and US victories. The editorials in April and May 1944 encouraged the Tibetans to pursue independence.[8] New Delhi further considered enlisting the support of the US by setting up a channel to facilitate better American public understanding of the situation in Tibet. It was advised that

> We would propose that further specific suggestions be put to the Tibetan Government to maintain a representative with the Government of India with headquarters at Gangtok and to take a suitable opportunity of sending an envoy with presents to the President of the United States of America in return for those which he sent to the Dalai Lama [via Dolan and Tolstoy] two years ago.[9]

In this spirit, American journalist A.T. Steele, who worked for *Life, Minneapolis Star Journal* and *Chicago Daily News,* visited Tibet and attended discussions between Basil Gould, the British Political Officer in Sikkim, and Shen Zonglian 沈宗濂, a former aide to Chiang Kai-shek who had been sent to Lhasa to operate a Chinese mission there.[10]

The Chinese did not sit idly by. Their countermeasures included increasing the number of broadcasts from Chongqing and Xichang 西昌 in Chinese-controlled Kham. Effort were also made to finance a weekly Tibetan-language newspaper (Chinese name: *Minxin Zhoubao* 民新周報) under a Guomindang party member of Tibetan descent, but this was hampered by geographical difficulties.

---

Lhasa. There had been an English school in Gyantse during the period from 1923 to 1926. Its closure can also be mainly attributed to the opposition of conservative elements in Tibet. See Alex McKay, *Tibet and the British Raj: the Frontier Cadre 1904–1947* (Richmond: Curzon Press, 1997), 115–117, 156.

**6** TBL, IOR/L/P&S/12/4217, telegram from External Affairs Department (GOI) to Zenodotia, London, April 3, 1944.

**7** IMH Archives, 11–29–26–00–015, *Ying dui Zang xuanchuan duli* [British Propaganda to Encourage Tibetans for Independence], telegram from Chinese Consulate-General at Calcutta to Foreign Ministry, December 10, 1944.

**8** IMH Archives, 11–29–26–00–015, *Ying dui Zang xuanchuan duli* [British Propaganda to Encourage Tibetans for Independence], letter from MAC to Foreign Ministry, August 6, 1944.

**9** TBL, IOR/L/P&S/12/4217, telegram from External Affairs Department (GOI) to Zenodotia, London, April 3, 1944.

**10** Lamb, *Tibet, China & India, 1914–1950,* 332.

Instead, radio receivers and loudspeakers in Lhasa and other major Tibetan cities were regarded as a more effective means of disseminating information.[11]

Meanwhile, it was estimated that over 50,000 boxes of British munitions had been conveyed to Tibet by mid-1944. The Chinese made a formal complaint about these arms sales, but the British did not respond. The Tibetan Regent, perhaps on British advice, sought to clarify that the arms and ammunition were purchased by Tibetan noblemen for their personal use.[12] For his part, Chiang Kai-shek ordered that some Chinese arms and ammunitions should be gifted to the Kashag. These included four "82" mortars (400 shells), eight Maxim guns (16,000 bullets) and 16 *zhongzheng* 中正 rifles (1,600 bullets).[13] Chiang wished to convey the message to Lhasa that while the British wished to profit from arms sales it could obtain arms and ammunition from the central government free of charge.

## 1.2 Wooing the Tibetan Elites

Sino-British competition was mainly targeted at influencing the attitude of the Tibetan elites. It was recorded in a Chinese source that Arthur J. Hopkinson, the new Political Officer in Sikkim who took over from Basil Gould in June 1945, gave senior Tibetan officials bullion, guns, radio sets and other gifts when he arrived in Lhasa. Twenty gold bricks were sent to each of the four Kalons (ministers) in the Kashag, each weighing 25 tolas (roughly 292.5 grams).[14] Shen Zonglian was, however, even more generous. An American-educated Chinese Buddhist, he was granted significant funds to undertake lobbying work in Tibet. He believed that "lobbying by money and valuable goods was equally as important as deterrents in dealings with the Tibetan elite."[15] In pursuit of this, Shen distributed over 100 valuable gifts, such as silk, gold and silver products, jewellery, lacquer wares, enamels, corals, gold watches, jade wares and tea-colour citrine glasses, to influential Tibetans. He also provided an electric train, a

---

**11** IMH Archives, 11–29–26–00–015, *Ying dui Zang xuanchuan duli* [British Propaganda to Encourage Tibetans for Independence], letter from MTAC to Foreign Ministry, September 4, 1944.
**12** Chen, *Kangzhanqianhou zhi Zhong-Ying Xizang jiaoshe*, 175.
**13** TBL, IOR/L/P&S/12/4194, telegram from Richardson (Lhasa) to External Affairs Department, New Delhi, July 14, 1944; ibid., 181.
**14** IMH Archives, 11–29–26–00–015, *Ying dui Zang xuanchuan duli* [British Propaganda to Encourage Tibetans for Independence], telegram from the Nationalist government to Foreign Ministry, November 8, 1945.
**15** Quoted and translated from Zhu, *Xizang zhuanshi*, 340.

mini-film projector and other high-grade toys for the young Dalai Lama.[16] Moreover, after he arrived in Lhasa in August 1944, lavish banquets were frequently held at his mansion to entertain Tibetan secular and religious elites. These parties usually lasted for a whole day, with luxurious food being served, and guests were entertained on occasion by Beijing Opera or Yunnan Opera performances.[17] Shen's Buddhist status made him particularly influential in Tibetan religious circles. Indeed, it was through his influence over the monks in the three largest temples of Lhasa that Shen managed to close the English school in Tibet.[18]

Through his wooing, Shen gradually established close connections with important figures in Tibet. His contact list contained the family members of the Fourteenth Dalai Lama, such as the Dalai Lama's father, Choekyong Tsering or Yabshi Kung (Qiquecairen 祁郤才仁), and his brother, Gyalo Thondup (Jialedunzhu 嘉樂頓珠). He also befriended Radreng Hutuktu, the ex-Regent, Tsarong Dazang Dramdul (Charong Dasangzhandui 擦絨·達桑占堆), the head of the Mint Bureau, and other notable Tibetan officials and lamas.[19] He further endeavoured to lobby Surkhang Dzasa[20] (Suokang Zhasa 索康·紮薩), the head of the Tibetan Foreign Bureau, who was an advisor to two Kalons, Kashopa Chogyal Nyima (Gaxue Qujinima 噶雪·曲吉尼瑪) and his son, Surkhang Wangchen Gelek (Suokang Wangqingele 索康·旺欽格勒).[21] Shen's strategy was to spark a reorganisation of the Tibetan cabinet in early 1946, with Surkhang Dzasa as the Silon (prime minister), a position second only to the Regent.[22] In 1945 – 46, some secret agreement seemed to have been reached between Surkhang and Shen and Surhkang committed himself to putting the Kashag under Chinese authority after he took power.[23]

Unsurprisingly, Shen's success in Lhasa gave rise to British anxiety. As a result, Basil Gould's retirement date was pushed back several times. The British-Indian government wanted to exploit the prestige of this senior Tibetan specialist to stiffen pro-British sentiments in Tibet and his presence in Lhasa in 1944 – 45

---

16 Ibid.

17 Ibid., 344.

18 Shengqi Liu, *Lasa jiushi [Memories regarding Lhasa]*, 2nd ed. (Beijing: Zhongguo Zangxue Chubanshe, 2014), 17. Liu Shengqi 柳陞祺 was then Shen Zonglian's English-language secretary.

19 Zhu, *Xizang zhuanshi*, 345.

20 "Dzasa" is his title and his full name is Surkhang Wangchen Tseten (Suokang Wangqincideng 索康·旺欽次登).

21 Zhu, *Xizang zhuanshi*, 354.

22 Lin, *Tibet and Nationalist China's Frontier*, 186.

23 Zhu, *Xizang zhuanshi*, 354 – 355.

seems to have had some effect.[24] Indeed, Surkhang Dzasa reassured him that "the Tibetan government could never trust any promises from China and that any settlement would have to be underwritten by a third party."[25] He confided to Gould that while he wanted to dispense with "suzerainty" since it implied that Tibet formed part of Chinese territory, he was also worried about difficulties in achieving this:

> the majority of educated Tibetans wanted to be independent of China, but that the uneducated, some of whom still think that the Manchu Emperors are in power, believe that what Lönchen Shatra proposed in 1914 is the best that Tibet can hope for.[26]

These discussions with Gould appear to have been at odds with his aforementioned agreement with Shen and it is not clear what Surkhang's true intentions were. However, it was apparent that Tibet, without guarantees of practical British support, was afraid to pursue independence or antagonise the Nationalist government. More military and diplomatic support was therefore required by the Kashag, particularly British facilitation of Tibet's attendance at the post-war peace conference.[27]

Gould made great efforts to persuade London to give more support to Tibet.[28] But it could not provide what the Tibetans needed. In an official reply to the Kashag, relayed by Gould, the following points were made:

> [a]. My Government earnestly desires that the autonomy of Tibet, including the right of Tibet to be in direct relations with the Government of India, should be preserved. You may feel quite certain that my Government would be ready to do all they can to help to secure these results by diplomatic means.

> [b]. I am not authorised to guarantee military support. For this there are three reasons. (a) It would, as you can see, be difficult for my Government to give such [a] guarantee in a matter which affects a country which is an ally of His Majesty's Government in the present great war. (b) My Government trust that neither Tibet nor China will allow an oc-

---

24 Lamb, *Tibet, China & India, 1914–1950*, 329. Basil Gould was a senior British Tibetan specialist who had served as the Political Officer at Sikkim (1913–1914, 1935–1937, 1937–1945) and in other appointments related to Tibet for decades and had established intimate relations with his Tibetan counterparts.
25 TBL, IOR/L/P&S/12/4194, points arising from discussion with the Tibetan Foreign Office, September 16, 1944.
26 Ibid.
27 TNA, FO 371/41589, telegram from Government of India to India Office, November 3, 1944.
28 TBL, IOR/L/P&S/12/4194, telegram from Sir. B.J. Gould to C.B. Duke (Deputy Secretary of the External Affairs Department, GOI), July 20, 1944.

casion for the use of force to arise. (c) My Government thinks that a satisfactory solution can be reached by peaceful means.

[c].  My Government points out that, because China is their ally in the war, His Majesty's Government are in a very favourable position to use their influence to bring about a peaceful settlement.

[d].  The presence of a Tibetan representative at the Peace Conference would not be appropriate because Tibet has not taken part in the war; and in any case it may be a long time before any formal Peace Conference takes place.[29]

This made the Kashag doubt whether the British supported Tibetan independence. As a result, there was a rethinking on the part of the Tibetans about direct negotiations with the Chinese. A Tibetan mission was sent to China in 1946, disguised by the Kashag as a delegation to celebrate the victory of the Allies so as not to antagonise the British.

## 1.3 Chinese Predicaments over Claiming the Assam Himalayas

British encroachment on South Tibet was particularly perturbing to the Chinese. The Tibetan Affairs Conference of the Nationalist government in December 1944 had agreed two initiatives:

1.  To establish a Frontier Agency which governs the defence of the Yunnan, Xikang, Tibet and Qinghai borders as a whole. Policemen for national boundary security should be stationed at the locations of high transport importance;
2.  To negotiate in due course with the British over these problems in line with the friendship principle.[30]

However, neither of these initiatives were carried through. The first, proposed by the Military Ordinance Ministry (Junlingbu 軍令部) of the MAC, was impractical due to a shortfall of Nationalist military forces, which were preoccupied with fighting the Japanese, as well as the semi-independent status of the Chinese border provinces adjacent to Tibet. The second was followed up perfunctorily by the Foreign Ministry. Gu Weijun merely made an enquiry in London in January 1945 about Gould's pressure on Tibet over the cession of Tawang. The British evaded the question on the pretext that Whitehall was not aware of his activities in Tibet

---

**29** TNA, FO 371/46121, aide-memoire to the Tibetan Foreign Office by B.J. Gould, December 4, 1944.
**30** IMH Archives, 11–29–26–00–024, *Guanyu Zang'an huiyi* [The Conference concerning Tibet], minutes of the Conference concerning Tibet, December 11, 1944.

as he was on a private trip there before his retirement with the purpose of saying farewell to his Tibetan friends.[31]

Two further practical difficulties contributed to Chinese ineffectiveness in intervening in the Indo-Tibetan frontier dispute. The first was Tibet's opposition to Chinese involvement. Shen sought to persuade the Kashag to let the Chinese central government negotiate on behalf of Tibet in the territorial disputes with the British, but without an effective outcome.[32] He then turned to Gould. In late 1944, Shen hinted that the Assam border problem could be settled easily once there was a comprehensive Anglo-Chinese arrangement over Tibet. Gould was unresponsive.[33] The Chinese ambassador to India made another attempt in 1947 when a Tibetan delegation was in New Delhi to attend the Asian Relations Conference. He told the Tibetans that "we the Chinese feel that it is better that you let us talk and handle this border issue. It will be more effective and more influential if we talk about it." The Tibetans rejected this "help" and added that if the border issue was raised, they would discuss it themselves; they had brought all the relevant documents.[34] The second difficulty was China's poor demographic and geographical knowledge of the Assam Himalayas. Tibet did not share any information and the formidable terrain made it hard for the Nationalists to make investigations themselves of the area. As a result, Chinese knowledge was mainly limited to the Lower Dzayül, which adjoined Chinese-controlled Kham and Yunnan, leaving the position in the middle and west portions, Loyül and Monyül, unclear.[35]

The Chinese were convinced that Britain's ultimate aim was to incorporate these lands into their domain. Even if unsuccessful, this could still benefit British interest in any negotiations over the Sino-Indian border arrangements.[36] In July 1946, Wang Shijie, who had taken over the Chinese Foreign Ministry from Song Ziwen the previous summer, made a formal complaint to Britain about

---

**31** IMH Archives, 11–29–26–00–015, *Ying dui Zang xuanchuan duli* [British Propaganda to Encourage Tibetans for Independence], telegram from Gu Weijun (London) to Foreign Ministry, January 15, 1945.

**32** Zhu, *Xizang zhuanshi*, 347–348.

**33** Lamb, *Tibet, China & India, 1914–1950*, 489.

**34** Goldstein, *Demise of the Lamaist State*, 563.

**35** A sketch map was drawn by August 14, 1945 but merely illustrated the situation from Ba'an 巴安 (now Batang) to Duorangong 奪然工 (Sadiya) (see map 11). See IMH Archives, 11–32–11–02–010, *Yingyin qinlue Xikang bianjing* [British-Indian Invasion to Border of Xikang], instruction and map of the route line from Ba'an to Duorangong, enclosed in the telegram from MAC to Foreign Ministry, August 13, 1945.

**36** IMH Archives, 11–32–11–02–010, *Yingyin qinlue Xikang bianjing* [British-Indian Invasion to Border of Xikang], letter from MAC to Foreign Ministry, August 23, 1944.

the latter's aggression in 1943 in South Tibet. London was in no mood to engage with the matter since it was now preparing to retreat from India; the issues concerning Tibet would be the Indian government's responsibility.[37] For its part, the Indian government did not intend to compromise the territory inherited from Britain. In April 1947 it replied to the Chinese government in the following terms:

> the Indian Embassy have the honour to state that the only activities in which the Government of India have been engaged in the area in question have been entirely restricted to the Indian side of the boundary between India and Tibet which has been accepted for over 30 years. The Government of India are also satisfied that no aircraft employed in connection with these activities have crossed the accepted border.[38]

A stalemate was thus reached. The crux of the Nationalist dilemma was the fact that their claims over South Tibet were exerted through their sovereignty over Tibet. If China's authority over Tibet was questioned so was its right over the Assam Himalayas. The key to unravelling the deadlock was therefore to restore Chinese sovereignty over Tibet, by either military or diplomatic means. Preoccupied with other domestic issues, and due to its military weakness, the Nationalist government chose to bring Tibet into its post-war political framework peacefully, by virtue of Chiang's new ethnic theory and the new "self-governance of high-degree" formula.[39]

## 2 Chinese Endeavours to Bring Tibet into Its Post-war Political Framework

### 2.1 The "High-degree" Self-governance Scheme

The formula of "high-degree" self-governance, or autonomy, was first raised by Shi Qingyang 石青陽, the head of the MTAC, in 1933. But at that time the nature of this concept was little different to the Han-Tibetan relations as defined in the Qing era. In 1942, the idea was revived by Wang Chonghui 王寵惠, Secretary-General of the Supreme National Defence Council of the Guomindang

---

37 IMH Archives, 11–32–11–02–010, *Yingyin qinlue Xikang bianjing* [British-Indian Invasion to Border of Xikang], letter for British Embassy (Nanjing) to Wang Shijie, December 5, 1946.
38 IMH Archives, 11–32–11–02–010, *Yingyin qinlue Xikang bianjing* [British-Indian Invasion to Border of Xikang], letter from Indian Embassy (Nanjing) to Chinese Foreign Ministry, April 11, 1947.
39 TBL, IOR/L/P&S/12/4194, speech addressed by Chiang at Supreme National Defence Council, August 25, 1945.

(SNDCG), under Chiang's instruction. This time, the concept of "high-degree" self-governance accorded full administrative powers to the Tibetan government but precluded those relating to defence and diplomacy, as well as some areas of transportation, economy, finance and education.[40]

This formula ostensibly conformed to the British notion of "autonomy" but also had its origins in Chinese domestic concerns. The first issue was Chiang Kai-shek's new interpretation of the relationship between Han and non-Han ethnic minorities. Since the late Qing era, the Chinese Han elite and other sinicised races were constantly spooked by secessionism prevailing in outlying territories. After the collapse of the Qing Empire in 1912, the doctrine of the five-nation republic (*wuzugonghe* 五族共和) was introduced by the Republican government with the purpose of unifying the different ethnic groups. According to this theory, the five major races – Hans, Manchus, Mongols, Huis (Chinese Muslims) and Tibetans – made up the Chinese nation, and harmony and equality were emphasised between these different ethnic groups. During the war Chiang Kai-shek developed a new version of this theory, the "racial theory of the Chinese nation" (*Zhonghuaminzu zongzulun* 中華民族宗族論). As outlined in Chiang's *China's Destiny*, the five major races were in fact five branches of the same ancestral origin.[41] This new interpretation aimed to underline the common identity of the different ethnic groups in China and to serve as a counterweight to the Japanese encouragement of secessionist movements among the Manchurian and Mongolian communities. The creation of "Manchukuo" ("Manzhouguo" 滿洲國) in 1932 and "Mengjiang" 蒙疆 in 1939, respectively covering Northeast China (Manchuria) and the middle part of Inner Mongolia, were products of this Japanese tactic. These two satellite regimes were supposed to have been formed in the name of the Manchu and Mongol people but in fact were under the sway of the Japanese Empire.

Another domestic consideration came from the Nationalists' own experience in handling the cases of Inner Mongolia and Xinjiang. The Nationalist programme in the early 1930s had been to grant local autonomy to all Chinese territories, without differentiating between ethnic minority areas and those inhabited mainly by the Han Chinese. Inner Mongolia, for example, was, as a first step,

---

**40** Zhu, *Xizang zhuanshi*, 351.

**41** Chiang, *China's Destiny and Chinese Economic Theory*, 39–40. The book was first published in 1943 in Chinese. It was widely believed that it was written by Tao Xisheng 陶希聖, one of Chiang's secretaries, rather Chiang himself. But, according to Chiang's diaries, the bulk of the book was personally written by Chiang.

demarcated into several provinces and each of these into several counties.[42] It thus had the same political structure as China proper. Self-governance was accorded to counties, leaving the major power centred at the provincial and central levels. But these reforms encountered huge opposition from Mongol aristocrats. The latter believed that the aim was to destroy their traditional domains and impair their inherited privileges since the rule of the Qing Empire over Mongolia had been indirect and through a system of *meng* 盟 (league of banners) and *qi* 旗 (banner).[43] As a compromise, the two systems co-existed for most of the 1930s and 1940s, but the Japanese manoeuvring over Inner Mongolian independence in the mid-1930s pressured the Nationalist regime into devolving more power to *meng* and *qi*.[44] As for Xinjiang, in mid-1942 the Guomindang regime resumed its rule by managing to shift the allegiance of its dictator, Sheng Shicai 盛世才, from the Soviet Union to the Guomindang. Nationalist troops subsequently marched into Xinjiang, but an ethnic rebellion soon broke out in 1944, incited and financed by the Soviet Union. The rebel Uyghurs came to control three counties out of ten in Xinjiang and established their own regime in Yining 伊寧 (Kuldja). Through a series of negotiations, the Yining regime was eventually incorporated into Nationalist Xinjiang. Yet it was allowed to co-organise the new provincial government within the Nationalist-administered counties. Furthermore, high-level officials were required to embody a diversity of ethnic backgrounds. Over half of the civil service positions were reserved for non-Han candidates, while in South Xinjiang the ratio rose to 70%.[45] It was against this backdrop that autonomy policies for Tibet emerged.

In August 1944, Wu Zhongxin 吳忠信, the head of the MTAC, submitted a "self-governance of high-degree" formula to Chiang Kai-shek under which Tibet was to be considered, along with Outer Mongolia. Apart from national de-

---

42 Inner Mongolia had already been partly sinicised in the late Qing time and thus had a much closer relationship with China proper. But for Outer Mongolia it was a different case.
43 In the Qing era the high authority of Beijing went down to every Mongol through a system of *meng* and *qi*. The *meng* was a steady alliance of several *qi*s and each *qi* consisted of several nomadic tribes. The system was based on a combination of assorted nomadic groups rather than territories, although each group had a relatively fixed pasture. Most chiefs and heads of *mengs* and *qis* were hereditary, and their political allegiance was dominated by a feudal hierarchy. Small headmen showed their loyalties to the bigger chiefs and ultimately to the Qing Emperor.
44 Qine Wu, "Zhonghuaminzu zongzulun yu Zhonghuaminguo de bianjiangzizhi shijian [The Theory of the Clan of Chinese Nation and the Practice of Local Autonomy in the Frontier of the Republic of China]," in *Chiang Kai-shek and the Formation of Modern China,* 2 vols., vol. 1, ed. Zijin Huang and Guangzhe Pan (Taipei: Institute of Modern History, Academia Sinica, 2013), 190–204.
45 Ibid., 204–211.

fence and foreign affairs, full self-governance was to be accorded to the respective "Special Autonomous Governments" of Outer Mongolia and Tibet. Yet the power to define the scope and nature of self-governance was to be retained by the central government. Furthermore, in the case of Tibet, the central government was not to intervene in religious affairs, but the theocratic system was to be abandoned, separating political power and legal jurisdiction from monastic authorities.[46] This "self-governance of high-degree" principle was confirmed by the Sixth National Guomindang Congress in May 1945. Thereafter, the Central Planning Board of the SNDCG, headed by Wang Shijie, was instructed to work it up into a feasible plan.[47]

Global politics, however, soon intervened. At the Yalta Conference of February 1945, Roosevelt agreed to Joseph Stalin's desire for an international recognition of Outer Mongolian independence and certain concessions in Northeast China in exchange for the Soviet promise of opening a second front in Northeast Asia as soon as possible after the end of war in Europe. The Chinese government was unaware of the full details until it became engaged in new treaty discussions with the Soviets in July 1945. In fear of the Soviet support of the Chinese Communist Party, the Guomindang regime was obliged to give way on the matter. This compromise triggered a chain reaction effect among the ethnic-minority communities in Inner Mongolia, Xinjiang and Tibet. As a countermeasure, the scope of "self-governance of high-degree" was widened. In an address to the SNDCG on August 24, 1945 Chiang Kai-shek stated that:

> As regards the political status of Tibet, the Sixth National Kuomintang Congress decided to grant it a very high degree of autonomy, to aid its political advancement and to improve the living conditions of the Tibetans. I solemnly declare that if the Tibetans should at this time express a wish for self-government our Government would, in conformity with our sincere tradition, accord it a very high degree of autonomy. If in the future they fulfil economic requirements for independence, the National Government will, as in the case of Outer Mongolia, help them to gain that status. But Tibet must give proof that it can consolidate its independent position.[48]

---

**46** IMH Archives, 11–29–26–00–021, *Meng Zang zhengzhi sheshi fang'an* [The Political Schemes for Mongolia and Tibet], the Scheme for Regaining Occupied *mengs* and *qis* in Mongolia and the Political Program for post-war Mongolia and Tibet formulated by Wu Zhongxin, August 27, 1944.

**47** The ceremonial head of the Central Planning Board was Chiang Kai-shek. Wang was the Chief Secretary. The main duty of this body was to formulate policies for post-war China. Wang Shijie was appointed Foreign Minister in July 1945 and Wu Dingchang 吳鼎昌 took over his position.

**48** TBL, IOR/L/P&S/12/4194, speech by Chiang at Supreme National Defence Council, August 25, 1945.

After consultations with the MTAC, the Foreign Ministry and the MAC, a five-point plan was released by the Central Planning Board in October 1945. Although it was more cautious about internal political reforms in Tibet, the gist of the doctrine was in line with Wu's proposal. It was proposed that high-ranking officials from the capital or Lhasa should be assigned to move the Tibetans towards a "self-governance of high-degree" and negotiations with the British should be kept on track.[49] In the meantime, Shen Zonglian was instructed to approach the Kashag with the new "self-governance of high-degree" formula.

## 2.2 The Tibetan Congratulatory Mission (Goodwill Mission) to India and China

The Chinese scheme of "self-governance of high-degree" was in conflict with Tibet's desire for *de facto* independence. Lhasa did not wish to give up either defence or diplomatic powers, let alone the centuries-old theocracy and privileges of the lamas in the temporal world.[50] The Kashag's stance was reflected in a nine-point communiqué prepared for Chang Kai-shek. This included the demand that China should return the Komondor (the area around Qinghai Lake) and Kham regions to Tibetan control and withdraw unwelcome Chinese influence in Lhasa. It was also reasserted that Tibet was an independent political entity and that the Sino-Tibetan relationship should be re-established on the traditional priest-patron basis. Notwithstanding, Shen Zonglian managed to arrange for the Tibetans to attend a conference in Nanjing to discuss these matters, along with Mongols, Uyghurs and other ethnic minorities. This conference turned out to be the Chinese National Assembly.[51] A Tibetan mission of eight members, camouflaged, as previously discussed, as a delegation to celebrate the victory of the Allied nations, was sent first to India and then on to China.

In India, the Tibetan mission was received by the American *Chargé d'affaires* to India under British facilitation. Letters from the Fourteenth Dalai Lama and the Kashag to President Truman were handed to the US representative, who later conveyed them to Washington. The letters were ceremonial but served as

---

**49** IMH Archives, 11–29–26–00–021, *Meng Zang zhengzhi sheshi fang'an* [The Political Schemes for Mongolia and Tibet], document of "five principles" from the Central Planning Board to Foreign Ministry, October 5, 1945.
**50** The Nationalists tended to attribute the Tibetan independence movement to its theocratic system. But, as discussed in the previous chapter, the most pro-Chinese elements were to be found in the three biggest monasteries in Lhasa, particularly the Drepung monastery.
**51** Lin, *Tibet and Nationalist China's Frontier*, 172.

a means of establishing a link between Tibet and America after the visit of Brooke Dolan II and Ilya Tolstoy in 1942–43. During the visit of the Tibetan mission, the Sino-British lobbying was continued in India. Arthur J. Hopkinson was nervous about the Tibetans attending the Chinese National Assembly, and when several members of the delegation developed pimples on their faces, hands and legs, this condition was intentionally exaggerated as a highly contagious disease to prevent them proceeding on to China. Yet after consulting a Chinese doctor the Tibetans were assured that they were being deceived by Hopkinson since the pimples were very common in people from cold climates who experienced the baking heat of India. After this incident, the Chinese arranged for the Tibetans to move into "China House" in Calcutta rather than the Great Eastern Hotel as prepared by the Indian government.[52]

The Tibetan delegation arrived in Nanjing on April 5, 1946 and were anxious to initiate discussions with their Chinese counterparts. The Chinese officials in the MTAC recognised the wide divergence between the views of Nanjing and Lhasa. Yet their ultimate aim was to coax the Tibetans into attending the Chinese National Assembly rather than engage them in any separate negotiations. To ensure their attendance at the Assembly, a significant sum of money was allocated to the mission for its daily expenses and a variety of tours were arranged, such as a summer visit to Beiping 北平 (Beijing). It was not until November that the mission finally submitted their nine-point communiqué to Chiang Kai-shek.[53]

The Tibetans found themselves being treated as formal delegates from Lhasa at the Assembly. Thubten Samphel (Tudengsangpei 土登桑培), one of the two heads of the mission, was accepted as a member of the Praesidium of the National Assembly. The Kashag instructed them to attend the meetings but refused them permission to vote or even clap their hands. The mission members were also told to block any resolutions concerning Tibet's status or other relevant questions.[54] After some ten days of plenary sessions, the eight members of the Tibetan mission, along with six Tibetan delegates from the Chinese-administrated areas (mainly from the camp of the Panchen Lama), were placed on the same committee with representatives from Inner Mongolia and Xikang to discuss relevant sections of the constitution. Two sections were related to Tibet. One was that "all the people of the countries whose delegates are present in this Assembly are subjects of the Chinese Nationalist Government." After asking for instructions from Lhasa, the mission was told not to accept this and to withdraw from

---

**52** Lamb, *Tibet, China & India, 1914–1950*, 495–497; Goldstein, *Demise of the Lamaist State*, 551–552; Lin, *Tibet and Nationalist China's Frontier*, 186.
**53** Zhu, *Xizang zhuanshi*, 355–356.
**54** Goldstein, *Demise of the Lamaist State*, 554–555.

the Assembly if the Chinese insisted on their agreement to this resolution.[55] The other section included Tibet and Inner Mongolia as parts of the Chinese polity that were to be permitted to continue to exercise "autonomy". The Tibetans protested about this to Bai Chongxi 白崇禧 (Pai Chung-hsi), the Defence Minister and head of the committee, but the latter responded that "China is like a corporation in that there are different components such as the Hans, Manchus, Mongols, Huis (Chinese Muslims) and Tibetans. Because China is the owner of these, there is nothing wrong with the Tibetan Question being included in the constitution."[56] In the end, the Tibetan mission, under the orders of the Kashag, departed for Shanghai in order to avoid signing the constitution.[57] Meanwhile, after the closure of the National Assembly, the Tibetans finally received a Chinese reply to their nine-point communiqué. The response was vague and merely paid lip service to Tibetan demands. It was suggested that the latter send another mission with decision-making authority to negotiate these questions. The MTAC hoped the new mission would attend the next Chinese National Assembly, but this never happened.[58]

In retrospect, the Tibetan mission, in Melvyn Goldstein's opinion, was "badly outmanoeuvred" by the Chinese.[59] But this does not adequately explain why the Tibetan delegates continued to attend the discussions in Nanjing after they had realised the genuine purpose of the Chinese. Perhaps the reason lay in the simultaneous conflict taking place in Tibet between Taktra Rimpoche,

---

**55** TBL, IOR/L/P&S/12/4226, memorandum from the British Mission in Lhasa to the political office in Sikkim, December 9, 1946.

**56** Qiang'eba Duoji'ouzhu, "Xizang difangzhengfu pai 'daibiaotuan weiwen Tongmengguo he chuxi Nanjing Guomindaibiaodahui' neimu [Secret Story of Congratulations of the Tibetan Delegation to the Victory of the Allied Nations and Its Attendance to the Chinese National Assembly]," in *Xizang wenshiziliao xuanji (2) [Selected Collections of Historical Sources Relating to Tibet (Vol. 2)]*, ed. Xizang zizhiqu zhengxie wenshiziliaoyanjiuhui (Beijing: Minzu Chubanshe, 1984), 8. The Tibetan complaints had some effect. Chiang decided to extract the word "local" from the original article that "Tibetan Local Self-Government System shall be safeguarded", despite the fact that some Chinese representatives were worried that this deletion might weaken the interpretation of China's relations with Tibet. See Hao Zhang, "Zhanhou de Xizang wenti yu 'Zhonghuaminguo xianfa' youguan tiaowen de zhiding ji shishi [The Post-war Tibet Question and the Stipulation and Implementation of the Relevant Clauses of the 1946 Constitution of the Republic of China]," *Shixue Yuekan* 2 (2015): 58–59.

**57** Goldstein, *Demise of the Lamaist State*, 558.

**58** Lin, *Tibet and Nationalist China's Frontier*, 173–174. But a Chinese source apparently informed that the Kashag had provided a list of those to attend the Second National Assembly. See Hao Zhang, "1946 nian Xizang difangdaibiao chuxi zhixianguoda wenti zhi tanxi [The Analysis of the Tibetan Attendance at the National Assembly in 1946]," *Shixue Jikan* 4 (July 2013): 79.

**59** Goldstein, *Demise of the Lamaist State*, 558.

the Regent at the time, and his predecessor, Radreng Hutuktu, which will be discussed below. In any event, the Tibetan attendance at the Chinese National Assembly was a tremendous propaganda victory for the Nationalists. It was the first official Tibetan visit to the Chinese capital since 1912. More importantly, Tibet's attendance at the Assembly seemingly integrated it into the Chinese constitutional framework. The 1946 Chinese Constitution was passed after the Tibetan delegates had been in attendance and Tibet was to be regarded as a self-governing district within Chinese territory.

## 3 The British Policy of a Tibetan Status Quo

### 3.1 The British Response to Chiang's "Self-governance of High-degree" Scheme

The British had been aware of the "self-governance of high-degree" formula in 1944. In a series of conversations with Shen Zonglian, Gould was told that "China was ready to recognise Tibetan [internal] autonomy but Tibet's external relations must be under Chinese control."[60] Shen also advised the British-Indian government that China wished to initiate formal Anglo-Chinese discussion over Tibetan affairs. He hinted that Tibet's mineral wealth could be exploited by British companies and Britain could retain many of the rights obtained under the Simla convention, and its resultant trade agreement, in a new Sino-British accord, in spite of the fact that the Chinese accepted neither of these two treaties as binding.[61] Indo-China trade via Tibet was also something to consider. The Chinese had begun to favour the building of a road from Jyekundo to Sikkim by way of Nagchuka (Naqu 那曲) and Lhasa, dubbed the Changlam Road (see map 13). This was different from either the Rima road entering India through Dzayül or the pack route which ran from Tachienlu to Sikkim via Chamdo and Lhasa.[62] Despite these inducements, the British government did not think it a good idea for the Tibetans to be excluded from such discussions.[63]

---

**60** TBL, IOR/L/P&S/12/4217, telegram from External Affairs Department (GOI) to Secretary of State for India, September 23, 1944.

**61** TBL, IOR/L/P&S/12/4217, telegram from External Affairs Department (GOI) to Secretary of State for India, October 4, 1944.

**62** TBL, IOR/L/P&S/12/4217, telegram from External Affairs Department (GOI) to Secretary of State for India, October 29, 1944.

**63** TBL, IOR/L/P&S/12/4217, minute by D.M. Cleary (IO), January 19, 1945, and letter from Sterndale Bennett (FO) to F.A.K. Harrison (IO), October 12, 1944.

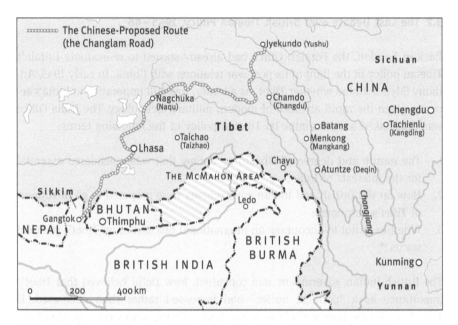

**Map 13:** The Proposed Changlam Road. © Zhaodong Wang & © Peter Palm, Berlin.

Meanwhile, in response to Chiang's statement on Tibetan autonomy in his speech to the SNDCG in August 1945, an agreement was reached between the Foreign Office and India Office over the way forward:

> If this autonomy is to be full autonomy, to include independence in external relations, we could, if the decision of the War Cabinet of 7th July 1943 is maintained, accept a recognition of Chinese suzerainty over Tibet, which we are not prepared to acknowledge unconditionally.[64]

The Foreign Office was, however, more sanguine about Chiang's scheme than the Indian government. It thus encouraged a China-Tibet rapprochement and believed that the Tibetan Congratulatory Mission to China afforded "an opportunity to raise this issue which is too good to miss."[65]

---

**64** TBL, IOR/L/P&S/12/4194, minute to the Under-Secretary of State (IO), September 10, 1945; TBL, IOR/L/P&S/12/4195A, memorandum by Cabinet Offices, titled "The Status of Tibet", October 26, 1945.
**65** TNA, FO 371/53613, memorandum by note by G.V. Kitson (FO), titled "The Status of Tibet", February 2, 1946.

## 3.2 The Last Debate over British Tibetan Policy, 1945–46

Back in London, the Foreign Office had already started to re-evaluate Britain's Tibetan policy in the light of its post-war relations with China. In early 1945, Anthony Eden doubted whether Tibet, as a buffer, was still imperative to India's security, given the rapid advance of modern military technology. The India Office was thus asked to re-examine its Tibetan policy in the following terms:

1. The nature and degree of Tibetan autonomy that was considered essential for the interests of India;
2. How far the British and Indian governments were prepared to go in support of Tibet's autonomy;
3. Whether or not to encourage an international discussion on Tibet's political status.[66]

The British-Indian government was consulted. New Delhi believed that Tibet's importance as a "political buffer" had increased rather than decreased. If Tibet fell into the hands of China or the Soviet Union, who were capable of developing it into an offensive base, India would be in great danger and large Indian military resources would have to be devoted to guarding the long northern boundary. For this reason, it argued for more British support for Tibet's *de facto* independence, not less.[67] It was also calculated that if Tibet was kept weak and secluded this would assist the Indians in asserting their treaty rights in the McMahon area. To add to this, it was contended that "Tibet should enjoy internal autonomy and the right to conduct direct foreign relations, recognising Chinese suzerainty by more or less ceremonial formalities."[68] Thus any attempts to explicitly define "suzerainty" should be continually avoided.[69] Furthermore, in consideration of the fact that the Indo-Tibetan border conflicts had already distanced Tibet from India, and that China's "self-governance of high-degree" scheme might persuade Tibet back to China, New Delhi believed it urgent to stiffen Tibet's resolve to resist Chinese ambitions. London was then advised openly to support Tibet's *de facto* independence:

---

**66** TBL, IOR/L/P&S/12/4194, letter from Foreign Office to India Office, January 2, 1945.

**67** TBL, IOR/L/P&S/12/4195A, copy of memorandum, titled "Policy Towards Tibet", from the Joint Secretary to GOI in the War Department to the External Affairs Department (GOI), May 18, 1945.

**68** TBL, IOR/L/P&S/12/4195A, telegram from the External Affairs Department (GOI) to India Office, September 19, 1945.

**69** TBL, IOR/L/P&S/12/4195A, minute by E.P. Donaldson (IO), December 29, 1945.

Support for Tibetan autonomy should be increased by the strongest and most outspoken diplomatic pressure at the present time (apart from other advantages, this would satisfy Tibet that our action in the McMahon areas is only our assertion of our own treaty rights and in no way designed to break up the state of Tibet), by publicity for the realities of Tibet's position over the last 33 years, and by the supply of munitions and equipment if sought by the Tibetan Government. ... There may be advantage in bringing the position of Tibet before the United Nations, but it is realised that the Chinese may be able to render such a course ineffective.[70]

For its part, the India Office felt it unnecessary to revise a policy reviewed just 18 months previously.[71] But it was alarmed about a possible compromise by the Foreign Office over Tibet in exchange for Chinese concessions over Hong Kong.[72] Its opinion was generally in line with that of the British-Indian government, although it believed the threat of the Soviet Union was overstated. It was in favour of British military training of Tibetan troops and sceptical about the impracticability of further military support.[73] The Joint Committee of the Chiefs of Staff and Air Headquarters was also consulted. Given that operations by aircraft in the high altitudes of Tibet were impracticable, and neither the Russians nor the Chinese had aircraft or rockets capable of operating from Tibet against India, its conclusion was that from a short-term point of view, there were neither "practicable means of aiding Tibet against a major enemy" nor "any real threat to India from that direction."[74]

As regards diplomatic support, the India Office was reluctant to place the case of Tibet on the stage of the newly formed United Nations because China, as a permanent member of the Security Council, had the right to "veto any proposal either that Tibet should become a member of the United Nations, or that her status should be discussed by the Council or Assembly." Nevertheless, propaganda should be employed to support the British position and a good means to facilitate this was to invite Steele, the American journalist mentioned earlier, to report on the Tibetan situation when the Tibetan Congratulatory Mission visited India. The India Office was convinced that such publicity would resonate with the current trend of autonomy in Asia, such as "China's recent agreement to

---

**70** TBL, IOR/L/P&S/12/4195A, telegram from the External Affairs Department (GOI) to India Office, September 19, 1945.

**71** TBL, IOR/L/P&S/12/4194, minute by India Office, January 24, 1945.

**72** Lamb, *Tibet, China & India, 1914–1950*, 334.

**73** TBL, IOR/L/P&S/12/4195A, minute by E.P. Donaldson (IO), December 29, 1945.

**74** TNA, FO 371/53615, explanatory note of Chiefs-of-Staff Committee by A. Macdonald, May 24, 1946.

grant self-governance to [Outer] Mongolia, the Wavell offer to India, and the Dutch and French offers to Netherlands East Indies and French Indo-China."[75]

The Far Eastern Department of the Foreign Office sided with the India Office, but its view did not entirely dominate its parent ministry. Ernest Bevin, the new Labour Foreign Secretary, objected, for example, to any connection between Britain's action in the McMahon area and appeasing Tibet. In his mind, "[the British] action in the McMahon area should be allowed to stand on its own merits." He also disagreed that the Chinese should be subjected to "the strongest and most outspoken diplomatic pressure" over Tibetan autonomy.[76] It was eventually agreed to leave the Tibetans to handle the autonomy issue with China themselves unless they called for British intervention.[77] As a result, the 1943 doctrine was to continue under Bevin. Yet the publicity machine was cranked up and propaganda materials were prepared for distribution in Britain, India and the United States.[78]

## 4 The Position in 1947

### 4.1 The Collapse of the Nationalist Intelligence Network and Its Influence in Tibet

By capitalising on the Tibetan mission that had attended the Chinese National Assembly, Chinese Nationalists achieved some propaganda gains. But two incidents greatly damaged Chinese leverage in Tibet.[79] The first was the exposure of the Tibetan Revolutionary Party (TRP) or Tibetan Improvement Party. This party, politically guided and subsidised by the Guomindang, had been established in 1939 with its base in the Indian border towns of Kalimpong and Darjeeling. Pan-

---

**75** TBL, IOR/L/P&S/12/4195A, minute by E.P. Donaldson (IO), December 29, 1945.

**76** TBL, IOR/L/P&S/12/4195B, letter from Secretary of State (IO) to External Affairs Department (GOI), March 5, 1946.

**77** TNA, FO 371/53613, minute by G.V. Kitson (FO), February 20, 1946.

**78** TBL, IOR/L/P&S/12/4195A, telegram from E.P. Donaldson (IO) to Foreign Office, January 8, 1946.

**79** As the Nationalist-administered sphere spread westward during the war, the Nationalists gradually established an intelligence network in the western border areas. Agents of the Investigation and Statistics Bureau of the Military Affairs Commission (ISBMAC) were first sent to Tibet in 1940. By 1944, there were 15 agents directly assigned by Chongqing and more than 10 agents recruited locally. Through this network, the Nationalists endeavoured to collect intelligence about Japanese activities in Tibet and cultivate pro-Chinese elements there, especially after the resignation of the Regent Radreng Hutuktu in 1941. See Zhu, *Xizang zhuanshi*, 362.

datsang Rapga (Bangda Raoga 邦達·饒嘎), the head of the party, was a devout believer in the political philosophy of Sun Yat-sen and wanted a similar revolution in Tibet to that which had taken place in China. His political blueprint for Tibet was in line with Nationalist ideals and theories. He was convinced that the Tibetan regime was ill-suited for the modern world and was determined to create an autonomous Tibetan Republic under the control of Republican China.[80] In 1943 Chiang Kai-shek agreed to support Rapga and ordered his secret agents working in Tibet and north India to collaborate with the TRP.[81] However, due to some TRP members' carelessness, the party's activities were discovered by the Indian police in early February 1946. Although the head of the Chinese commission in Delhi had warned Rapga of the imminence of a raid, some documents relating to the party were not destroyed, including a party agreement and several letters written by Rapga to Chinese officials.[82] Soon afterwards, Rapga and his associates were accused of spying and revolutionary activities against the Tibetan government. After claiming Chinese nationality, Rapga was eventually deported to China in July 1946. His party, and the party's intelligence network set up with Nationalist cooperation in Tibet, were yet virtually destroyed.[83]

Further damage was caused by the Guomindang's non-intervention in the Radreng-Taktra clash of 1945–1947. The Radreng Hutuktu was the Regent of Tibet from 1934 to 1941. During his reign, he had facilitated two Chinese official missions to Lhasa, led by Huang Musong 黃慕松 in 1934 and Wu Zhongxin in 1940. In contrast to Taktra Rimpoche's frosty attitude towards the Chinese, Radreng Hutuktu was widely believed by Chinese officials in the 1940s to be a pro-Chinese figure. He resigned in January 1941, for reasons which are unclear, and invited Taktra Rimpoche to assume power.[84] But apparently there was an agreement between them that Taktra Rimpoche would return the regency to Radreng Hutuktu three years later. However, when the time arrived, Taktra Rimpoche broke his promise.[85] In response, Radreng Hutuktu renewed his secret con-

---

**80** Goldstein, *Demise of the Lamaist State*, 450.

**81** Lin, *Tibet and Nationalist China's Frontier*, 185.

**82** Goldstein, *Demise of the Lamaist State*, 458.

**83** Lin, *Tibet and Nationalist China's Frontier*, 191.

**84** In his letter to the Nationalist government, he explained that his life was in great danger since he was "prophesied" ill by God's will. See ibid., 185. Yet, in Melvyn Goldstein's opinion, the main reason was due to his sexual misdeeds. See Goldstein, *Demise of the Lamaist State*, 357–360.

**85** It was worth stating that three years on Radreng Hutuktu still had influence in Tibet. According to a British official, when Radreng Hutuktu arrived in Lhasa on December 3, 1944 he was greeted as the returning Regent. A reception tent was erected one mile east of Lhasa, where all the government officials, including the Dalai Lama's father and other family members, gath-

tacts with the Guomindang and sought Chinese military support to overthrow Taktra Rimpoche's regency. Yet the presence of the Tibetan mission – which was monopolised by the Taktra faction – at the Chinese National Assembly held Chiang Kai-shek back from an endorsement of Radreng Hutuktu. This explains why Taktra Rimpoche chose not to withdraw his mission from Nanjing despite having realised Chinese intentions.[86] There was also a divergence of opinion between the MTAC and the Investigation and Statistics Bureau of the MAC (ISBMAC) over whether it was worth endorsing the Radreng faction.[87] As a result, an opportunity to intervene in the conflict was lost. Before long, Radreng Hutuktu was arrested after a failed assassination attempt on Taktra Rimpoche and a rebellion mounted by his backers was suppressed by Taktra Rimpoche's troops. In May 1947 Radreng Hutuktu died in prison and his associates were subsequently purged.

Radreng Hutuktu's demise dealt a severe blow to Chinese prestige in Tibet. He was not only a member of the Guomindang Central Executive Committee, but also his religious title of "Hutuktu"[88] had been granted by the Guomindang regime. The failure of China to support him resulted in much damage to pro-Chinese elements in Tibet and a notable decline of confidence in the Nationalist regime.[89] The morale of the Nationalist officials in Tibet also suffered a serious setback. A Chinese intelligence agent noted that

> Concerning the incident, so far only three telegrams have been dispatched from the Central Government, to which the Tibetan Government turns a deaf ear and pays no heed. Han Chinese sympathisers here are rather dissatisfied with the disposition of the Central Government and think that it will be a thousand times more difficult to solve the Tibetan question after Taktra entirely eradicates Tibetan pro-Han factions. [It is my perception that] the Cen-

---

ered to greet him. See TBL, IOR/L/P&S/12/4201, memorandum from Major G. Sherriff (Lhasa) to Political Officer in Sikkim, December 3, 1944.

86 The selection of the members of the Tibetan mission was monopolised by the Taktra faction and no delegates came from the three largest monasteries in Lhasa which were more inclined towards Radreng Hutuktu. See Zhang, "1946 nian Xizang difangdaibiao chuxi zhixianguoda wenti zhi tanxi [The Analysis of the Tibetan Attendance at the National Assembly in 1946]," 67–68; Lin, *Tibet and Nationalist China's Frontier*, 189.

87 Radreng Hutuktu, in the MTAC's eyes, was an incompetent and greedy politician. However, Wei Long 魏龍, an agent of the ISBMAC in Tibet, believed that only if Radreng Hutuktu and the pro-Chinese faction took power could Tibetan issues be solved. See Chen, *Kangzhanqianhou zhi Zhong-Ying Xizang jiaoshe*, 377–383; Zhu, *Xizang zhuanshi*, 354–368; Lin, *Tibet and Nationalist China's Frontier*, 190.

88 "Hutuktu" was a Tibetan religious title only inferior to the Dalai Lama and the Panchen Lama which no longer exists today.

89 Zhu, *Xizang zhuanshi*, 372.

tral Government is willing neither to dispatch troops [to introduce calm] nor to make any political gestures [with respect to the incident]. Thus, in the future, government officials working in Tibet will no longer be able to secure a foothold here.[90]

To compound the damage, the Kashag found evidence of Radreng Hutuktu's collusion with the Nationalist leaders through correspondence found in his hermitage. As a consequence, the Chinese underground network, which linked Radreng Hutuktu and the Nationalist regime, was exposed and suffered a devastating blow. As a result, it was even more difficult for Chinese officials and agents to secure a hold on Tibet. In this sense, echoing Lin Hsiao-ting's argument, the Nationalist efforts in Tibet had failed by 1947.[91] Two years later, the Kashag expelled all Chinese officials from Tibet on the pretext that their allegiance might be to the new Chinese Communist regime.[92]

## 4.2 The British Retreat from India and the Stance of the New Indian Government

An interim government of India was formed on September 2, 1946 to assist the transition of British India to independence. During the process of the transfer of power, the Foreign Office ceased operations in Tibet.[93] It was resolved that India, not Pakistan, would inherit all British treaties and arrangements with Tibet. On August 15, 1947 India became an independent state and took control of the British bureaucracy in Tibet. But it soon transpired that no Indian officials were suitable to replace Hugh Richardson, the then head of the British Mission in Lhasa. Richardson thus served as the head of the Indian body in Tibet until his retirement in 1950.[94]

In general, the new Indian government assumed the British mindset over Tibet. Despite Chinese protests, Jawaharlal Nehru invited the Tibetans to join a

---

**90** Intelligence report from Jin Da (Guomindang intelligence officer in Lhasa) to Chiang Kai-shek, May 5, 1947, quoted in Lin, *Tibet and Nationalist China's Frontier*, 182.

**91** Ibid., 197.

**92** This included all secret agents of the Nationalist intelligence system in Tibet. See Zhu, *Xizang zhuanshi*, 378.

**93** Lamb, *Tibet, China & India, 1914–1950*, 508.

**94** After departing from India, British Tibetan specialists planned a goodwill mission to Lhasa in 1949 in order to maintain an Anglo-Tibetan relationship, but this never materialised. See Qiang Zhai, *The Dragon, the Lion, and the Eagle: Chinese-British-American Relations, 1949–1958* (Kent, Ohio and London, England: The Kent State University Press, 1994), 49; ibid., "Tibet and Chinese-British-American Relations in the Early 1950s," 38.

semi-official conference of Asian countries in the spring of 1947 as a way to increase Tibet's international profile. During the conference, the Tibetan delegation was eager to discuss the abolition of the former British privileges in Tibet and Indo-Tibetan border issues. Yet Nehru brushed aside these demands, informing them that neither Tibet's political status nor the border question would be on the agenda. In the end the Tibetan delegation only talked with the Indian foreign minister about Indo-Tibetan trade issues.[95]

Nonetheless, there were some fresh developments. The new Indian leaders, to some degree, downgraded the geopolitical importance of an independent Tibet to India. They showed more interest in close cooperation between India and China, which they believed was essential to the development of a new pan-Asian order. Lesly A.C. Fry, an official in the UK High Commission in New Delhi, observed in the autumn of 1947 that

> The conditions in which India's well-being may be assured and the full evolution be achieved of her inherent capacity to emerge as a potent but benevolent force in world affairs – particularly in Asia – demand not merely the development of internal unity and strength but also the maintenance of friendly relations with her neighbours. To prejudice her relations with so important a power as China by aggressive support of unqualified Tibetan independence is therefore a policy with few attractions. It follows that, while the Government of India are glad to recognise and wish to see Tibetan autonomy maintained, they are not prepared to do more than encourage this in a friendly manner and are certainly not disposed to take any initiative which might bring India into conflict with China on this issue. The attitude which they propose to adopt may best be described as that of a benevolent spectator, ready at all times – should opportunity occur – to use their good offices to further a mutually satisfactory settlement between China and Tibet.[96]

The Indians also showed a more conciliatory attitude over their special rights in Tibet inherited from the British. In a discussion, K.P.S. Menon, the Indian ambassador, told Ye Gongchao 葉公超, the Chinese Vice Foreign Minister, that

> I can assure you that our approach to the Tibetan problem will not be the same as that of the British Government in the past. We have no designs on Tibet, but we would like to keep those privileges in Tibet which are not detrimental to China.[97]

---

**95** Goldstein, *Demise of the Lamaist State*, 562.
**96** TNA, FO 371/63943, memorandum by L.A.C. Fry (Office of the United Kingdom High Commissioner in India), titled "Summary of Indo-Tibetan Relations", enclosed in the letter from L. Fry to E.P. Donaldson (Commonwealth Relations Office), November 7, 1947.
**97** IMH Archives, 11–29–26–00–022, *Feichu Zhong-Ying guanyu Xizang zhi bupingdengtiaoyue* [The Abolition of Unequal Treaties between China and Britain regarding Tibet], minutes of talks between the India Ambassador Mr. Menon and Vice-Minister Yeh, October 15, 1947.

Furthermore, while the Indian government stood by the McMahon Line, it was "prepared to discuss in a friendly way with China and Tibet any rectification of the frontier that might be urged on reasonable grounds by any of the parties to the abortive Simla Conference of 1914."[98]

Later, these new Indian attitudes developed into a two-faceted policy. Indian support for Tibetan independence consistently dimmed. When, for example, the PRC acted to incorporate Tibet into Chinese territory in 1950–1951, the Indian government refused munitions requested by Tibet and a proposal on the part of the United States to intervene in Tibetan affairs. Nehru insisted that Indo-Chinese relations outweighed any obligations inherited from Britain over Tibet.[99] Yet Indian opinion on the disputed Indo-Tibetan border continued to harden. This, to some degree, contributed to an Indo-Chinese border war which broke out in 1962.

# 5 The Tibetan Inheritance

## 5.1 Attempts to Abolish Former British Privileges in Tibet

In 1946–1947, when the transition of power was underway in the sub-continent of South Asia, Chinese diplomats did not want to miss this opportunity to resolve the Tibetan issue in their favour. They deliberated on how to terminate British privileges by applying new Sino-British treaties to Tibet. In their opinion, Britain should still be treated as the negotiating party, rather than India, and it was in the Chinese national interest not to recognise India's inheritance of British treaty rights and obligations. In addition, this would avoid offending Pakistan who did not inherit any British privileges in Tibet and the Himalayas.[100] But the British would not agree to any such renegotiation.

At the same time, there were Tibetans who advocated a new arrangement with the independent Indian regime. They argued for the replacement of the terms "Chinese suzerainty" and "Tibetan autonomy" in the Simla convention with that of "Tibetan Independence"; for India's retrocession of the areas with

---

**98** TBL, IOR/L/P&S/12/4210, telegram from L.A.C. Fry (New Delhi) to the Political Officer in Sikkim, April 8, 1947.

**99** Zhai, "Tibet and Chinese-British-American Relations in the Early 1950s," 42–48.

**100** IMH Archives, 11–29–26–00–022, *Feichu Zhong-Ying guanyu Xizang zhi bupingdengtiaoyue* [The Abolition of Unequal Treaties between China and Britain regarding Tibet], memorandum of the abolition of unequal treaties regarding Tibet by the Chinese Foreign Ministry, August 10, 1948.

Tibetans in a majority, such as the Assam Himalayas, Darjeeling and Sikkim; and for the revision of trade and economic agreements which were detrimental to Tibetan interests. The Tibetan Foreign Bureau indicated that it was in no mood to abide by the existing treaties inherited by the Indian government:

> Regarding our request [for the] return of excluded Tibetan territories gradually included into India and regarding trade relations affecting [the] general economic welfare of Tibet, we have discussed the matter with [the] Government of India when India was under British administration. And it becomes [necessary that] [the] Government of Tibet must continue the negotiations with new Government of India in [the] near future. Hence, we hope that His Majesty's Government will also support and help us in achieving our desire.[101]

However, this policy, as Melvyn Goldstein has argued, carried a risk of alienating Tibet from the new Indian authorities: "a development that could leave Tibet without any formal relations with a foreign country."[102] The Kashag thus made no real attempt to terminate British treaty rights in Tibet. On August 15, 1947, the British Mission in Lhasa, and three trade agencies in Gyantse, Gartok (Gadake 噶大克) and Yatung, simply replaced the Union Jack with the Indian national flag. The Tibetan government continued their interactions with these new Indian bodies as if no changes had occurred.[103]

Chen Xizhang 陳錫璋, the new head of the Chinese Mission in Lhasa, also made some effort to persuade the Kashag to abolish British treaty rights but with no success.[104] Chinese diplomats then hoped to persuade India to give up its inherited rights. Yet, like the British, the Indians stood by the Simla convention. It was thus concluded by the Chinese Foreign Ministry that "given that the Tibet question had now developed into a complex political puzzle, it was impossible to solve the problem merely through applying the theory of international law", and that "the only way forward was to create an entirely new treaty framework with India to replace all previous treaties regarding Tibet."[105] Howev-

---

**101** These Tibetan words are cited in a British document, see TNA, FO 371/63943, letter from L. Fry to E.P. Donaldson (Commonwealth Relations Office), November 7, 1947.

**102** Goldstein, *Demise of the Lamaist State*, 566–567.

**103** Ibid., 568–569.

**104** IMH Archives, 11–29–26–00–022, *Feichu Zhong-Ying guanyu Xizang zhi bupingdeng-tiaoyue* [The Abolition of Unequal Treaties between China and Britain regarding Tibet], letter from the MTAC to Foreign Ministry, March 27, 1948.

**105** IMH Archives, 11–29–26–00–022, *Feichu Zhong-Ying guanyu Xizang zhi bupingdeng-tiaoyue* [The Abolition of Unequal Treaties between China and Britain regarding Tibet], report by the Foreign Ministry in regard to the Ensuing Sino-Indian Discussion over Tibetan Issues, April 1948.

er, these plans were not put into practice since the Guomindang regime soon collapsed in mainland China.

## 5.2 The Tibetan Trade Mission and a New US Attitude towards Tibet

In mid-1947, the Kashag decided to send a trade mission to India, China, the United States and Britain, with the purpose of improving Tibetan trade with these countries. By this means, the Tibetan government sought to reduce Indian domination over its economic links with the outside world, especially the United States.[106] At this time, about 80% of Tibetan goods, mainly wool, furs, yak tails and musk, were exported to India, while 20% went to China. In turn, half of the wool exported to India was sold to the United States by American agents; the rest was resold by Indians abroad.[107] However, the Tibetans could not obtain foreign currency through these transactions. Foreign currency, mainly US dollars, was accrued in the Reserve Bank of India and the latter credited the equivalent in rupees to Tibetan merchants.[108] It was estimated that the income of these exports via India was about two or three million US dollars per annum.[109] Furthermore, during wartime, Tibetan goods were affected by Indian restrictions on imports and exports to and from India. These resulted in difficulties for Tibet over importing foreign goods and caused a rapid decline in gold and silver reserves.[110] Therefore, the trade mission also sought to purchase gold and silver in the United States and Britain in order to back up Tibetan paper money. More generally, the Kashag saw the mission as an excellent opportunity to increase Tibetan international visibility and to publicise its status as a country independent of China. With this purpose in mind, the Kashag issued official passports to members of the mission and instructed them to use them when travelling.[111]

---

[106] Telegram from the Ambassador in China (Stuart) to the Secretary of State, February 24, 1948, US Department of State, ed., *Foreign Relations of the United States Diplomatic Papers, 1948, the Fast East, China*, vol. VII (Washington, D.C.: Government Printing Office, 1973), 757–758. (hereafter *FRUS: 1948 China*)

[107] Goldstein, *Demise of the Lamaist State*, 571–572.

[108] Letter from Tibetan Trade Commission (Shakabpa) to the Secretary of State, August 7, 1948, US Department of State, ed., *FRUS: 1948 China*, 776–778.

[109] Goldstein, *Demise of the Lamaist State*, 572.

[110] Letter from Tibetan Trade Commission (Shakabpa) to the Secretary of State, August 7, 1948, US Department of State, ed., FRUS: 1948 China, 776–778.

[111] Goldstein, *Demise of the Lamaist State*, 572.

The trade mission, led by the chief of the Tibetan Financial Department, Tsipon Shakabpa (Xiageba Wangqiudedan 夏格巴·旺秋德丹), arrived in India in November 1947. Its negotiations with the Indian government soon ran aground. New Delhi refused to allot dollars for the purchase of gold and silver in the United States,[112] or hand over dollars which the Tibetans earned from the export of their products.[113] Nehru reiterated that Indo-Tibetan trade agreements would not be reviewed until Tibet formally agreed that India was the legal inheritor of the treaty rights and obligations of British India. Under such pressure, the Tibetan government announced its acceptance of Nehru's India as the successor to British India on June 11, 1948.[114]

Whilst in India, the Tibetan Trade Mission called upon the US Embassy in advance of a possible visit to the US. American diplomats in India were keen to build a close relationship with Tibet. One and a half years earlier, when the Tibetan Goodwill Mission had visited the American Embassy in India, the *Chargé d'affaires*, George R. Merrell, had attempted to get Washington to recognise the geopolitical importance of Tibet to the US. In his opinion, Tibet was in "a position of inestimable strategic importance both ideologically and geographically" to resist Soviet expansionism.[115] However, the Department of State at that time had no interest in arranging a trip to Lhasa. It was remarked that

> [The War Department's Plans and Operations Division] have stated informally that ... that area would not readily lend itself to development as a base from which ground, air or rocket operations could be effectively launched. Moreover, from the standpoint of Sino-American relations, the Department [of State] feels that no useful purpose would be served at the present time by action likely to raise the question of our official attitude with respect to the status of Tibet.[116]

Now the new American *Chargé d'affaires*, Howard Donovan, tried again to interest Washington in the value of Tibet. He suggested that the Americans give a cordial reception to a visit from the Tibetan Trade Mission. Although the importance

---

**112** Telegram from *Chargé d'affaires* in India (Donovan) to the Secretary of State, January 5, 1948, US Department of State, ed., *FRUS: 1948 China*, 755–756.
**113** Letter from Tibetan Trade Commission (Shakabpa) to the Secretary of State, August 7, 1948, ibid., 776–778.
**114** Goldstein, *Demise of the Lamaist State*, 574–575.
**115** Telegram from *Chargé d'affaires* in India (Merrell) to the Secretary of State, January 13, 1947, US Department of State, ed., *Foreign Relations of the United States Diplomatic Papers, 1947, the Fast East, China*, vol. VII (Washington, D.C.: Government Printing Office, 1972), 588–592. (hereafter *FRUS: 1947 China*)
**116** Telegram from Acting Secretary of State to the *Chargé d'affaires* in India (Merrell), April 14, 1947, ibid., 594.

of such a visit from a purely commercial point of view would be small, it was thought that a friendly reception in the United States might reinforce the US's relations with a government which controlled a large strategic area bordering on territory where Soviet influence was widespread.[117]

Behind the scenes, the British were quietly influencing the Americans. In a private letter to T. Eliot Weil, Second Secretary at the American Embassy, Arthur J. Hopkinson, the former British and now Indian Political Officer in Sikkim, repeatedly emphasised the importance of Tibet as "a vast island in Asia still apparently unaffected by Soviet influence" and "an area which in the future might prove extremely useful for military operations."[118] The Americans at this time had few channels to understand Tibetan affairs except through these British experts.[119] Indeed, the American Embassy in India was so short of personnel able to translate from Tibetan into English that it was obliged to ask for the help of the External Affairs Department of the Indian government. The letters from Lhasa to President Truman in 1946 had to be translated by Hopkinson's personal assistant, Lobzang Tsering (Luosangcairen 洛桑才仁).[120]

The State Department duly agreed to receive the Tibetan Trade Mission but was worried about the mission becoming a source of friction between the United States and China. It thus instructed its embassy in India that if the Tibetans insisted on using their own passports, American visas would be issued using Form 257, which was the standard procedure when applicants presented the passport of a government the US did not recognise.[121] The embassy preferred that the Tibetans apply for such visas in China.[122] The British were also wary of the status

---

**117** Telegram from *Chargé d'affaires* (Donovan) to the Secretary of State, December 30, 1947, ibid., 604–606.
**118** Telegram from the Ambassador in India (Grady) to the Secretary of State, August 21, 1947, ibid., 598–600. The excerpts of Hopkinson's letter were attached to the telegram.
**119** Telegram from the Ambassador in India (Grady) to the Secretary of State, November 21, 1947, ibid., 602–603.
**120** Here refers to the three letters simultaneously sent from the Dalai Lama, the Tibetan Regent and the Tibetan Cabinet Ministers. They were handed to the American representative when the Tibetan Goodwill Mission visited the American Embassy in India in May 1946. See Telegram from *Chargé d'affaires* in India (Merrell) to the Secretary of State, January 13, 1947, ibid., 588–592.
**121** Telegram from the Acting Secretary of State to the *Chargé d'affaires in India* (Donovan), December 26, 1947, ibid., 604.
**122** Memorandum by the Secretary of Embassy in India (Weil), December 30, 1947, ibid., 606–608.

of any trade visit to London and suggested the Tibetans might come in a private capacity rather than as an official mission.[123]

In January 1948, the Tibetan Trade Mission reached China through Hong Kong. The Guomindang government was naturally apprehensive about any Tibetan visit to America and Britain as an independent nation. At an inter-departmental conference in the Foreign Ministry, the Economic Ministry and the MTAC it was agreed that

    a.  The best solution was to persuade the Tibetan Trade Mission back to Tibet;

    b.  If the Tibetan mission insisted on travelling to the United States and Britain, it was desired that the Economic Ministry covertly organised a trade commission of non-governmental nature to mingle with the Tibetan mission;

    c.  The Tibetan Trade Mission should be accompanied by Chinese officials of both the Foreign Ministry and the Economic Ministry. This should be conducted in the name of the MTAC.[124]

The Chinese Foreign Ministry subsequently issued Chinese passports to the members of the Tibetan mission, but the Tibetans refused to use them. Under strong Chinese diplomatic pressure, neither Britain nor the United States declared themselves willing to issue visas on Tibetan passports – at least openly. The Guomindang also refused to issue exit permits on the basis of Tibetan passports. In the meantime, to help persuade the mission to go back to Tibet, Chiang Kai-shek apparently directed that $3,000,000 worth of silk and $750,000 worth of tea be shipped to Tibet. In return, the Tibetan mission had to promise not to proceed to the United States and Britain and to return to Tibet via Hong Kong. But Chiang was outmanoeuvred. The Tibetan mission managed to obtain US visas using their Tibetan passports from the American Consulate in Hong Kong and sold the silk in the colony whilst there.[125] Meanwhile, when the mis-

---

**123** United States Foreign Relations, 693.003 Tibet/1–1348, memorandum of conversation, January 13, 1948. Quoted in Goldstein, *Demise of the Lamaist State*, 576.

**124** AH Archives, 020–012600–0006, *Xizang pai shangwukaochatuan fu Ying Mei dengguo huodong I* [Activities of the Tibetan Trade Mission to Britain and the United States, etc., vol. 1], minutes of the Conference about the Tibetan Trade Mission to Britain and the United States, March 5, 1948.

**125** According to Goldstein's research, there is an alternative version of events: that Shakabpa never received any tea but purchased the silk on the open market with Chiang Kai-shek's permission. This silk was then allegedly shipped to Tibet by an Indian company. See Goldstein, *Demise of the Lamaist State*, 582–583.

sion was in Nanjing, the members had gained British visas using their own passports.[126]

The Chinese government was incensed and lodged a strong diplomatic protest with the United States and Britain.[127] In response, the US State Department misinformed the Chinese that the US visas had been issued on Form 257 and not on Tibetan passports.[128] The Chinese kept up their pressure. They requested that any Tibetan activities in the United States be arranged through the Chinese Embassy and the Tibetans be accompanied by a Chinese official. This was refused by the State Department on the grounds that the members of the Tibetan mission were considered "businessmen on a purely commercial basis."[129] As it turned out, the mission was received by the Department of Commerce, rather than the State Department, and due to Chinese insistence that the Chinese ambassador must be present at meetings, the Tibetan mission gave up the chance of an audience with President Truman. Yet it did succeed in having private talks with George Marshall, the Secretary of State. Marshall agreed to a transaction of 50,000 ounces of gold bullion (roughly 2 million US dollars) but the Tibetans did not have sufficient US dollars to purchase them.[130] The Tibetans then asked for a temporary loan, yet without success.[131] In the end, they received just 250,000 dollars from their Indian bank accounts, with Nehru's approval, when they returned to India.[132]

Nonetheless, the offer of gold by the United States signalled a change in attitude since the Gold Reserve Act of 1934 forbade any sale of gold except to foreign governments in the name of their Central Banks.[133] Although Marshall emphasised that any gold sale would not affect "the continuation of [the US

---

126 TNA, FO 371/70042, letter from the British ambassador, Nanjing, to the Foreign Office, May 19, 1948.

127 Memorandum of Telephone Conversation, by the Chief of the Division of Chinese Affairs (Sprouse), July 12, 1948, US Department of State, ed., *FRUS: 1948 China*, 759–760.

128 Goldstein, *Demise of the Lamaist State*, 586.

129 Memorandum of Telephone Conversation, by the Chief of the Division of Chinese Affairs (Sprouse), July 19, 1948, US Department of State, ed., *FRUS: 1948 China*, 763–764.

130 Letter from Secretary of State to the Leader of the Tibetan Trade Mission (Shakabpa), in New York, August 27, 1948, ibid., 779–780.

131 Letter from the Leader of the Tibetan Trade Mission (Shakabpa) to the Secretary of State, August 31, 1948, ibid., 783–784. Letter from Secretary of State to the Leader of the Tibetan Trade Mission (Shakabpa), in New York, September 27, 1948, ibid., 785–786.

132 Goldstein, *Demise of the Lamaist State*, 606.

133 Memorandum of the Second Secretary of Embassy in India (Weil), December 30, 1947, US Department of State, ed., *FRUS: 1947 China*, 606–608.

government's] recognition of China's *de jure* sovereignty over Tibet",[134] it was not hard to see that American attitudes towards Tibet had altered. This was reflected in a message to the US ambassador in China from the Secretary of State:

> The US government [has] no intention of acting in a manner to call into question China's *de jure* sovereignty over Tibet. ... [The] Chinese government should appreciate that the fact that it exerts no *de facto* authority over Tibet is [the] root cause of [the] situation. ... The press [shows] considerable interest in Tibetans' visit, and if it should become known that their intended call on the President was frustrated by the Chinese government, [I] believe that [the] press would make [the] most of [the] situation to China's disadvantage. Such [a] story might also be raised in light of self-determination which is [a] popular concept among [the] American people.[135]

Staying longer than expected in the US, the Tibetans found that their British visas had expired. The British decided to grant them new visas on affidavits as they had explained to the Chinese that their previously issuing of visas on Tibetan passports was a "technical error".[136] This infuriated the Tibetans, who threatened to cancel their trip to London. As a compromise, the British quietly extended their previous visas. As in the United States, the Board of Trade, rather than the Foreign Office, was designated as the mission's host in London. The Chinese likewise requested that the Tibetan meetings with the British must be approved by the Chinese Embassy. Yet they were refused on the grounds that the mission was "informal and confined to matters of trade and commerce."[137] The Prime Minister, as well as the Lord Chamberlain, met the Tibetans, although these meetings were low profile.[138] Little was achieved commercially. The Tibetans' request for pound sterling was rebuffed by London and after three weeks the mission returned to India in December 1948.

Despite the lack of concrete achievements, the trade mission demonstrated the limits of Chinese control over Tibet. More significantly, Lhasa's links with the United States had now been re-established after Dolan and Tolstoy's visit to Lhasa in 1943 and, in view of the forthcoming collapse of the Nationalist regime in mainland China and the western confrontation of Communist expansion in

---

**134** Letter from Secretary of State to the Secretary of the Treasury (Snyder), August 27, 1948, ibid., ed., *FRUS: 1948 China*, 780.

**135** Telegram from the Secretary of State to the Ambassador in China (Stuart), July 28, 1948, ibid., 767–768.

**136** TNA, FO 371/70044, letter from Shakabpa to the British ambassador in the United States, October 9, 1948. Quoted in Goldstein, *Demise of the Lamaist State*, 598–599.

**137** TNA, FO 371/70044, report on meeting of the British Foreign Office with an officer of the Chinese Embassy in England, October 28, 1948. Quoted in ibid., 601.

**138** Ibid., 604.

Asia, the United States started to readjust its policy towards Tibet. In April 1949 the State Department commented that

a. It is believed to be clearly to our advantage under any circumstances to have Tibet as a friend if possible. We should accordingly maintain a friendly attitude toward Tibet in ways short of giving China cause for offense. We should encourage so far as feasible Tibet's orientation toward the West rather than towards the East.

b. For the present we should avoid giving the impression of any alteration in our position toward Chinese authority over Tibet such as for example steps which would clearly indicate that we regard Tibet as independent, etc. ... We should however keep our policy as flexible as possible by avoiding reference to China's sovereignty or suzerainty unless references are clearly called for and by informing China of our proposed moves in connection with Tibet, rather than asking China's consent for them.[139]

The memorandum also pointed out that if China fell to the Communists it would be better to deal with Tibet as an independent nation rather than as part of a Communist Chinese state.

## Conclusion

In retrospect, the Guomindang regime had two opportunities to exert its sovereignty over Tibet after the war, either by exploiting the Radreng-Taktra clash or by preventing India from inheriting British treaty rights and obligations regarding Tibet. In neither case were they able to exploit the situation to their advantage. Moreover, trapped in the civil war from 1946 onwards, the Guomindang had neither the political will nor the material strength to incorporate central Tibet into Chinese territory. Indeed, the Nationalist elements were expelled entirely by the Kashag from central Tibet in July 1949, even before their collapse in mainland China.[140] Nonetheless, Chiang's "high-degree autonomy" formula had provided a prototype used later by the Chinese Communist Party (CCP) for its Seventeen-Point Agreement with Lhasa and its governing Tibet before 1959. Meanwhile, railing against Chiang's vision of a mono-ethnonational

---

139 USFR, 693.0031 Tibet/1–849, memorandum on US policy toward Tibet sent by the Far Eastern Affairs Department of the State Department to the Chinese Affairs Division of the State Department, April 12, 1949. Quoted in ibid., 609.

140 The young and powerless Panchen Lama was recognised in the Qinghai province by the Guomindang regime in 1949, but he and his associates chose to stake their political futures on the new Chinese Communist regime. See Lin, *Tibet and Nationalist China's Frontier*, 200–201.

China, the CCP gradually devised a new ethnonational theory: a unified multinational China comprised of 56 distinct but equal ethnic groups.[141]

As for the British, their approach to the Tibetan question, as well as their Tibetan bureaucratic structures, were largely inherited by independent India, although Nehru expressed more interest in strengthening India's relations with China and paid less attention to supporting Tibet's *de facto* independence. Meanwhile, India continued to claim the Assam Himalayas (the McMahon area) and thus gave rise to continuing conflicts with the PRC over this border. This even led to a war in 1962. Further afield, the British helped to facilitate US-Tibetan connections which contributed to a revaluation of the importance of Tibet in the climate of the Cold War. This exacerbated the confrontation between Communism and Capitalism in Asia.[142]

The Sino-British rivalry in the region thereafter played out along two lines: the border conflicts between India and the PRC; and arguments between the United States and China over the independence of Tibet and, after 1959, the legitimacy of the exiled Tibetan government in India.[143] Even today, these conflicts are still in evidence. The White House frequently interacts with, and subsidises, the Fourteenth Dalai Lama and his exiled government, using the latter as leverage to counter the PRC.[144] In addition, the border disputes continue to hinder collaboration between China and India and might perhaps even trigger a new future war between these two countries.[145] In this sense, Sino-British relations over Tibet during the period 1941–1949 still influence present-day international politics.

---

141 Thomas S. Mullaney, *Coming to Terms with the Nation: Ethnic Classification in Modern China* (Berkeley, Los Angeles and London: University of California Press, 2011), 2–34. The policy evolved gradually and the number 56 was reached in 1981 (after the 1953–1954 registration the number was 39).

142 Hsiao-ting Lin, *Taihai, Lengzhan, Chiang Kai-shek: jiemi dang'an zhong xiaoshi de Taiwan shi 1949–1988 [Taiwan Strait, Cold War and Chiang Kai-shek: the Re-discovered Taiwan History in the Unclassified Archives 1949–1988]* (Taipei: Lianjing Chuban Shiye Gufen Youxian Gongsi, 2015), 202–237.

143 India also sometimes jointed the conflict, for example, to accommodate the Fourteenth Dalai Lama and his exiled regime but it is not an important player compared to the United States.

144 John Kenneth Knaus, "Official Policies and Covert Programs: The U.S. State Department, the CIA, and the Tibetan Resistance," *Journal of Cold War Studies* 5, no. 3 (2003): 54–79; ibid., *Beyond Shangri-la: America and Tibet's Move into the Twenty-First Century* (Durham, North Carolina: Duke University Press, 2012), 287–300.

145 Recent crises are the military standoff in Doklam (Donglang 洞朗) in mid-2017 and skirmishes at locations along the Actual Control Line between Indian Ladakh and Chinese Aksai Chin (Akesaiqin 阿克赛钦) in 2020.

# Conclusion

Negotiations over post-war Sino-British relations had an encouraging start in 1942–43 but by 1949 had failed to reach a satisfactory settlement. Not only did they fail to rebuild the two countries' commercial relations on an equal and reciprocal basis (as in the aborted commercial treaty) but they also did not terminate the British informal empire in China (with regard to Hong Kong and Tibet). The reasons for the failure were complex, encompassing both internal and external factors, including the powerful influence of the United States.

China's Second World War was not just a war against Japan but also a struggle to end western imperialism. In the latter effort, the Nationalist government was not a passive player. Through successfully playing the American card during the 1943 Sino-British treaty negotiations, it achieved the abolition of British imperialistic privileges in China that went far beyond the scope of extraterritoriality. Washington was also persuaded tacitly to confirm Chinese sovereignty over Hong Kong and Tibet. However, this policy of using the influence of the United States as a diplomatic tool had its limits. Chongqing gradually lost the support of Washington as China's value in defeating Japan decreased. The Sino-American relationship cooled, especially after the quarrels over surrendering Chiang Kaishek's command of Chinese forces to General Joseph Stilwell and after the death of Roosevelt.

Nevertheless, the Nationalists' enthusiasm for a comprehensive commercial treaty with Britain was not dampened. Yet developments in post-war China hampered Chongqing. After the war, the forces of Chinese nationalism and anti-imperialism increased and spiralled out of the Guomindang's control. The public opinion was pushing for a more aggressive approach to protect Chinese national industries by exerting stricter controls and tariffs over foreign economic activities and goods. In this context, the Sino-American commercial treaty, although signed on an equal and reciprocal basis, was criticised fiercely by the Chinese press for opening up the Chinese market widely to American companies and goods. Nationalist officials thus became more cautious about Sino-British terms on the grounds that the drafts of the two treaties were derived from the same model. In the end, negotiations failed to secure a treaty.

In relation to Hong Kong, the Guomindang was forced to pursue a postponement policy. In view of the decision of the Truman administration to side with Britain over its reoccupation of the colony after the Japanese capitulation, coupled with the distraction of the internal power struggle with the CCP, the Nationalists opted to shelve their claims over Hong Kong until after the cessation of domestic hostilities. In the interim, the Chinese public lobbied the Guomindang

https://doi.org/10.1515/9783110706659-014

regime hard to restore the colony. As a compromise, Nanjing acquiesced in the local authorities' attempt to restore Chinese administration over the Kowloon Walled City. This, however, caused arguments with the British over the sovereignty of the city. The dispute even deteriorated into violence in Guangzhou, where angry Chinese demonstrators burned the British consulate and other buildings. Yet the Nationalist regime, in the face of British intransigence, was not in a position to exert any leverage over the issue of Hong Kong.

As for Tibet, the crux of its separation from China derived largely from Lhasa's mistrust of the Chinese and pursuit of independence. Because it was preoccupied with the war against Japan and domestic conflicts with the CCP, the Nationalist government had no power to reconquer Tibet militarily. Its efforts thus focused on regaining the Kashag's political loyalty while seeking to win international recognition for its sovereignty over the region. The Pacific War, nevertheless, provided a chance for Chongqing to extend its solid influence into Tibet. With this aim in mind, the Nationalist regime actively exploited the exigency of a new supply line by way of Tibet, either building a vehicle road via eastern Tibet or supervising the animal pack shipment within Tibet's central areas. Though these plans were thwarted by Lhasa's strong opposition, which the British government fomented, some progress was made. British recognition of Chinese suzerainty over Tibet was retained although on the condition of a Chinese commitment to Tibet's autonomy. Under Sino-American pressure, British military support for Tibet's independence wavered and the Anglo-Tibetan relationship became strained over the British occupation of the disputed McMahon area. Yet the Guomindang failed to seize key opportunities. At the Chinese National Assembly of 1946, the Chinese scheme of "high-degree self-governance" was rebuffed by the Tibetans. Meanwhile, its intelligence network was exposed, and its influence waned, due to the government's incompetent manipulation of the power struggle inside Lhasa. There was then a wider transition of power in South Asia that the Guomindang could have capitalised on. However, independent India inherited British treaty rights in Tibet without hindrance, including the disputed McMahon area. The matter of Tibet was left unresolved.

For their part, the British strived to resist the collapse of their "informal empire" in China. London's concessions over the 1943 treaty, particularly over its rights beyond the scope of extraterritoriality, were made in the face of Sino-American pressure. Thereafter, Whitehall sought to regain some of the abandoned rights through the commercial treaty, such as foreign trade and navigation in Chinese coastal and inland waters. Meanwhile, departmental differences and conflicting opinions between London and parts of the Empire made it hard for the British to draft a treaty. For example, the Home Office and the Ministry of Labour disagreed with the Board of Trade's proposal to loosen controls over Chinese entry to the UK in exchange for reciprocal treatment for British business-men visiting and living in China. At the imperial level, India, Burma and Malaya refused to allow their national interests to be traded as part of the treaty. In the face of the uncertain political environment in China, the line of "no treaty is bet-ter than a bad treaty" came to dominate and treaty discussions with China were suspended.

In the case of Hong Kong, British concessions over the 1943 treaty increased the importance of the colony. To London, it was a key entrepôt for British mer-chants to trade with China and East Asia, especially in view of the absence of extraterritoriality and the chaos in the region after the war. In the meantime, while it was acknowledged that Hong Kong's return to China was inevitable in the longer term, views in Whitehall over the way forward were not in unison. The Foreign Office and the Colonial Office were at odds over the details of such a compromise and, in particular, over the open assertion of British inten-tions to remain in Hong Kong. In any event, with China in the throes of civil war, and the Americans now supporting Britain's return to Hong Kong, there was little incentive to compromise.

As regards Tibet, London's strategic preference was for a weak Tibet separat-ed from China. But this was at odds with maintaining China as a stable and friendly trading partner. The war complicated matters further. As George Sherriff, the head of the British mission in Lhasa, had pointed out, Britain could not be an ally both to China and to Tibet. As things transpired, London and New Delhi made capital out of anti-Chinese sentiments among the Tibetans and their desire for independence while at the same time controlling the disputed McMahon area in anticipation that the Chinese would have a stronger claim over these territo-ries if they regained Tibet. The independent Indian government then maintained the British position and there was a deadlock. London, meanwhile, endeavoured to sway the Americans over Tibet. It was no longer a remote and barren land but a forward position in the struggle to contain and confront Communist expansion. Against this background, the United States came to be the principal supporter of the Tibetan independence movement and financed the Fourteenth Dalai Lama and his exiled Tibetan government in India after 1959.[1]

The transition of power in London in mid-1945 brought a far-reaching change to Britain's domestic policy – the establishment of the National Health Service (NHS), more emphasis on economic planning, and nationalisation of the Bank of England, the coal mines, electricity and railways, etc. – but little to its foreign

---

**1** Lin, *Taihai, Lengzhan, Chiang Kai-shek*, 219–237; Knaus, "Official Policies and Covert Pro-grams," 54–79; ibid., *Beyond Shangri-la*, 287–300.

policy. Ernest Bevin, the Labour Foreign Secretary, proceeded with the position of the wartime coalition government on major questions, particularly in respect of the Empire.[2] Continuity can also be observed in the stable personnel of the Foreign Office – there developed a notable agreement on the retention of the Empire and regarding the Soviet Union as the central threat to Britain.[3] The independence of India was not expected but situations there were potentially too explosive to maintain British rule. Yet Clement Attlee managed to keep it, including Pakistan and Ceylon, within the British Commonwealth of Nations. After retreating from India, London became more prudent about its dealings with Lhasa despite that some Tibet experts frequently called for a resumption of intimate bilateral connections. It was worried that any British intervention in Tibet might not only offend China but also independent India as the latter was now the inheritor of British treaty arrangements with Tibet.[4] The continual emphasis on the Empire also gave rise to an uncompromising attitude of the Labour government over Hong Kong and British imperial remnants in other places of China, mainly economic rights and assets in Shanghai and along the Yangtse River. This coincided with diminished external forces of pressing Britain to come to terms with China, either from the Nationalist or American side. Furthermore, the Labour government was faced with tremendous international challenges after the war, many of which had a more significant impact on the country. For example, the negotiations in Geneva and Havana, primarily with the United States, aimed to lay the foundation of international trade and commerce for decades to come.[5] In this context, commercial treaty negotiations with China could wait. Chinese issues reverted to being a low priority on Britain's global agenda.

The United States was indeed an important external factor in the Sino-British relationship during the Pacific War owing to the fact that both China and Britain relied heavily on the former's resources to win the war. Although Roosevelt had a strong aversion to European colonialism, his primary wartime aim was not the dismemberment of the British Empire but the accumulation of enough allied strength to defeat the Axis powers. He thus embraced the abolition of the US's colonial rights in China to help maintain Chongqing's role in the fight against the Japanese, while conceding Chinese interests in Northeast

---

2 John Saville, *The Politics of Continuity: British Foreign Policy and the Labour Government, 1945–46* (London and New York: Verso, 1993), 93–100.
3 Ibid., 12.
4 TBL, IOR/L/P&S/12/4195B, telegram from J.S.H. Shattock to L.B. Walsh-Atkins (CO), October 27, 1948; Zhai, "Tibet and Chinese-British-American Relations in the Early 1950s," 38.
5 Douglas A. Irwin, Petros C. Mavroidis and Alan O. Sykes, *The Genesis of the GATT* (Cambridge and New York: Cambridge University Press, 2008), 114–122.

China to the Soviet Union in exchange for the latter's early participation in the war, so saving American money and lives. In the Sino-British case, the desire to end the informal British empire in China frequently gave way to a more urgent need: namely the maintenance of harmonious allied relationships. Roosevelt endeavoured to find a middle way that satisfied both China and Britain, but with little success. After his death, Washington seemingly lost interest in the Sino-British disputes. Yet, as the Cold War heated up, the United States re-engaged and US hostility towards Communist China in large part influenced Britain's approach to Communist China after 1950.[6]

The tortuous Sino-British negotiations over their post-war relations, and their many twists and turns, in some ways constituted a miniature version of the wider decolonisation process in East Asia during and immediately after the Pacific War. As part of this process, the United States and Britain gradually arrived at a compromise over colonial issues, laying down goals of decolonisation that ranged from "national independence" to "self-governance".[7] This paved the way for a return of European colonial powers to Southeast Asia after the war, as well as the restoration of British jurisdiction over Hong Kong. In parallel with this was the decline of China's position in the post-war Asian Pacific order led by the United States. According to Roosevelt's plan, China would be regarded as one of four major peace-keeping powers and the US's most important ally in East Asia. The Nationalist government therefore actively participated in the foundation of a series of international organisations such as the United Nations, the World Bank, the International Monetary Fund (IMF) and the General Agreement on Tariffs and Trade (GATT). However, with the demise of the Nationalist regime in mainland China, and the demilitarisation and democratisation of Japan under American occupation, Japan replaced China as the pillar of the US Asia-Pacific strategy. The Chinese, after the CCP came to power, in turn joined the Communist camp and allied with the Soviet Union. As the climate of the Cold War spread across the Asian Pacific, the Sino-British relationship was played out in most disadvantageous circumstances.

The ensuing diplomatic relationship between Britain and China was subject to the constraints of their respective Capitalist and Communist blocs during the Cold War. Consequently, although London recognised the Beijing government as early as January 1950 and exchanged *chargés d'affaires* in 1954 soon after the end of the Korean War, Sino-British relations were not fully re-established

---

6 Tang, *Britain's Encounter with Revolutionary China, 1949–54*, 194–196; Chi-Kwan Mark, *The Everyday Cold War: Britain and China, 1950–1972* (London: Bloomsbury Academic, 2017), 5.
7 Thorne, *Allies of a Kind*, 599.

until the United States and China reconciled in the 1970s. In 1974, Britain and China eventually restored their relationship at the ambassadorial level. In the meantime, British efforts to create new trade relations with Beijing were held back by US hostility towards Communist China.[8] The involvement of British forces in the Korean War further drove the Chinese government to impose harsher laws against British companies in China and at length expelled them from the country. Thus, whilst London signed a "Correspondence regarding British Trade in China" agreement with Beijing in 1952, it was not until 1978 that Sino-British commercial relations were rebuilt under the EEC (the European Economic Community)-PRC framework.[9] In 1987 disputes over British historical properties in China were finally solved.[10]

Concurrently, the western blockade of Communist China during the Cold War increased the importance of Hong Kong as a vehicle for mainland China to access western technology and commodities. In the view of Beijing, it was thus more advantageous to China for Britain to retain control of Hong Kong. Largely because of this, the status quo of Hong Kong continued for another half a century until British retrocession was ensured through the Sino-British joint declaration of 1984.[11] Hong Kong finally returned to China in 1997. Yet Sino-British conflicts over Hong Kong have not come to a halt. British colonial legacy is still evident in the city and so is London's intervention in Hong Kong politics. Before the handover, democratic reform of allowing the election of half the Legislative Council by universal suffrage was hurriedly and unilaterally launched and a system of British National Overseas (BNO) citizenship was introduced to retain Hong Kong people's link with the UK.[12] Since 1997, Britain has

---

**8** Tang, *Britain's Encounter with Revolutionary China, 1949–54*, 193–195.

**9** The trade agreement between the European Economic Community and the People's Republic of China was signed in April 1978.

**10** UK Treaties Online, "Agreement between the Government of the United Kingdom of Great Britain and Northern Ireland and the Government of the People's Republic of China concerning the Settlement of Mutual Historical Property Claims with Exchange of Notes concerning Shares, Bonds and other Securities," 1987, https://treaties.fco.gov.uk/awweb/pdfopener?md=1&did=68428 (accessed March 23, 2019).

**11** Ibid., "Joint Declaration of the Government of the United Kingdom of Great Britain and Northern Ireland and the Government of the People's Republic of China on the Question of Hong Kong with Annexes," 1984, https://treaties.fco.gov.uk/awweb/pdfopener?md=1&did=68291 (accessed March 23, 2019).

**12** The BNO passport allows its holder to enter and stay in the UK for six months without a visa. In 2021, a new system of the BNO visa, in place from January 31, extends the time to five years but with an application. After five years, the visa holders are eligible for "indefinite leave to remain", and finally, after a further 12 months, they are able to apply for British citizenship.

continued to play an active role in supporting and sympathising with Hong Kong's pro-democracy, perhaps also pro-independence, movements. Colonial flags are at times waved at public assemblies by a few Hong Kong citizens to express their nostalgic sentiments for British rule but, more significantly, strong disgruntlement against the central government. London and Beijing frequently accuse each other of breaching the terms of agreements under which the colony was handed back to China in 1997. London claims that it has a historical and moral commitment to Hong Kong people while this, in Beijing's eyes, is tantamount to interfering in China's domestic issues since the sovereignty of Hong Kong has already been transferred. It can be anticipated that these quarrels will hardly abate, at least unless 2047 when the Chinese commitment of "one country and two systems" reaches its 50-years limit. Yet, against the new background of the China-United States confrontation in the twenty-first century, things may become even worse.

The matter of Tibet evolved and became more complicated. After the Beijing government incorporated Tibet into its Chinese territories in 1951, the issue evolved into two strands. The United States, in place of Britain, became the primary sponsor of the Tibetan independence movement, using Tibet as leverage to contain Chinese Communism. US presidents, as well as other western leaders, frequently meet with the Fourteenth Dalai Lama, the head of the exiled Tibetan government in India. At the same time, independent India, while coming to acknowledge Chinese sovereignty over Tibet,[13] continued to control the McMahon area – or in the Chinese vernacular Zangnan 藏南 (South Tibet) – and claimed it as part of Indian territories.[14] This territorial dispute, along with other border disputes between China and India, gave rise to an Indo-Sino border war in 1962 and

---

[13] India confirmed China's sovereignty over Tibet in the Sino-Indian agreement of 1954 and the Chinese government reciprocated with an acknowledgement of many Indian privileges in relation to trade and transport with Tibet. After that, the Indian government's stance became a little ambiguous over the matter. India gave asylum to the 14th Dalai Lama and his followers in 1959. After the border war with China in 1962, India increased its support for the Dalai Lama and allowed him to set up a so-called government-in-exile in Dharamshala, although the Indian government has never acknowledged this. It was until 1988 that Rajiv Gandhi, the Indian Prime minister, reiterated that it was India's long-standing and consistent policy that Tibet was an autonomous region of China. In 2003, Prime Minister Atal Bihari Vajpayee confirmed that the Tibetan Autonomous Region is part of the territory of the People's Republic of China.

[14] In 1951, the Indian troops invaded the Tawang area and removed the Tibetan administration there. During the Sino-Indian war of 1962, Tawang fell briefly under Chinese control, but China voluntarily withdrew its troops at the end of the war. After that, India resumed its administration over Tawang but China has never relinquished its claims on the McMahon area with Tawang included.

numerous subsequent military clashes. Recent examples are the Indo-Chinese border standoff at Doklam in 2017 and skirmishes in 2020 at several locations along the line between Indian Ladakh and Chinese Aksai Chin.

In this regard, the legacy of British imperialism in Tibet and Hong Kong still affects East and South Asia today.

# Bibliography

## Unpublished archival sources

The National Archives, Kew, London:
    papers of the Foreign Office, the Colonial Office, the Board of Trade and the Cabinet
    Office.
The British Library, London:
    papers of the India Office and the British-Indian government.
Archives of the Institute of Modern History, Academia Sinica, Taipei:
    papers of three departments at the Chinese Ministry of Foreign Affairs (外交部): the
    departments of Asia and the Pacific (亞太司), the department of Europe (歐洲司) and
    the department of laws and regulations (條例司).
Academia Historica, Taipei:
    papers relating to the Chinese Ministry of Foreign Affairs and the Committee of
    Mongolian and Tibetan Affairs (蒙藏委員會): https://ahonline.drnh.gov.tw/index.php?
    act=Archive.
The Official Report of All Parliamentary Debates (Hansard):
    https://hansard.parliament.uk.
UK Treaties Online:
    https://treaties.fco.gov.uk/responsive/app/consolidatedSearch/.
The Churchill Archive (online):
    http://www.churchillarchive.com.
Foreign Office Files for China, Adam Matthew Digital:
    http://www.archivesdirect.amdigital.co.uk/FO_China.

## Thesis

Whewell, Emily. "British Extraterritoriality in China: the Legal System, Functions of Criminal
    Jurisdiction, and Its Challenges, 1833–1943." PhD diss., University of Leicester, 2015.

## Published archival sources

Ashton, S.R., G. Bennett, and K. Hamilton, eds. *Documents on British Policy Overseas, Series
    1, Volume 8: Britain and China 1945–1950.* London: Routledge, 2002.
Best, Anthony, ed. *British Documents on Foreign Affairs – Reports and Papers from the
    Foreign Office Confidential Print, Part III (1940–1945), Series E (Asia),* 8 vols, Vol. 5–6,
    Bethesda, MD: University Publications of America, 1997.
Chinese Ministry of Information, ed. *China Handbook 1937–1945: A Comprehensive Survey of
    Major Developments in China in Eight Years of War.* New York: The Macmillan Company,
    1947.

https://doi.org/10.1515/9783110706659-015

Reynolds, David, and Vladimir Pechatnov, eds. *The Kremlin Letters: Stalin's Wartime Correspondence with Churchill and Roosevelt.* New Haven and London: Yale University Press, 2018.

The Historical Committee of the Central Committee of the Guomindang, ed. *Zhonghuaminguo zhongyao shiliao chubian (Kangrizhanzheng shiqi): zhanshi waijiao [Preliminary Compilation of Important Historical Sources Relating to the Republic of China (the Period of the Anti-Japanese War): Wartime Diplomacy].* 3 vols. Vol. 3. Taipei: Zhongyang Wenwu Gongyingshe, 1981.

The Second Historical Archives of China, ed. *Zhonghuaminguoshi dang'anziliao huibian (Nanjing Guominzhengfu 1927–1949): Waijiao [Compilation of Archives and Documents Relating to the Republic of China (the Nanjing Nationalist Government 1927–1949): Diplomacy].* 3 vols. Vol. 1–3. Nanjing: Jiangsu Guji Chubanshe, 1997.

The Second Historical Archives of China, ed. *Zhongguo Guomindang Zhongyangzhixingweiyuanhui changwuweiyuanhui huiyijilu [The Minutes of Meetings of the Standing Committee of the Central Executive Committee of the Nationalist Party].* 44 vols. Vol. 1–44. Guilin: Guangxi Normal University Press, 2000.

The Second Historical Archives of China, ed. *Zhonghuaminguoshi shiliao changbian [The Extensive Compilation of Historical Archives Relating to History of the Republic of China].* 70 vols. Vol. 57–70. Nanjing: Nanjing University Press, 1993.

US Department of State, ed. *Foreign Relations of the United States Diplomatic Papers, 1942, China.* Washington, D.C.: Government Printing Office, 1960.

US Department of State, ed. *Foreign Relations of the United States Diplomatic Papers, 1943, China.* Washington, D.C.: Government Printing Office, 1957.

US Department of State, ed. *Foreign Relations of the United States Diplomatic Papers, 1943, the Conference at Cairo and Tehran.* Washington, D.C.: Government Printing Office, 1961.

US Department of State, ed. *Foreign Relations of the United States Diplomatic Papers, 1945, the Far East, China.* Vol. VII. Washington, D.C.: Government Printing Office, 1969.

US Department of State, ed. *Foreign Relations of the United States Diplomatic Papers, 1947, the Fast East, China.* Vol. VII. Washington, D.C.: Government Printing Office, 1972.

US Department of State, ed. *Foreign Relations of the United States Diplomatic Papers, 1948, the Fast East, China.* Vol. VII. Washington, D.C.: Government Printing Office, 1973.

Wang, Jianlang, ed. *Zhonghuaminguoshiqi waijiaowenxian huibian 1911–1949 [Compilation of Foreign Archives in the Chinese Republican Era 1911–1949].* 10 vols. Vol. 6–10. Beijing: Zhonghua Shuju, 2015.

Wu, Jingping, and Tai-chun Kuo, eds. *Fengyunjihui: Song Ziwen yu waiguorenshi huitan jilu, 1940–1949 [T. V. Soong: Selected Minutes of Meetings with Foreign Leaders, 1940–1949 (the Chinese and English Languages)].* Shanghai: Fudan University Press, 2010.

Zhou, En'lai. "Our Foreign Policy and Mission (30th April 1952)." In *Zhou En'lai Xuanji [Selected Collection of Zhou En'lai's Speeches and Articles],* edited by Zhonggong Zhongyang Wenxian Yanjiushi Bianji Weiyuanhui, 85–89. Beijing: Renmin Chubanshe, 1984.

## UK treaties

UK Treaties Online. "Agreement between the Government of the United Kingdom of Great Britain and Northern Ireland and the Government of the People's Republic of China concerning the Settlement of Mutual Historical Property Claims with Exchange of Notes concerning Shares, Bonds and other Securities." https://treaties.fco.gov.uk/awweb/pdfopener?md=1&did=68428.

UK Treaties Online. "Convention between the United Kingdom and China respecting an Extension of Hong Kong Territory." http://foto.archivalware.co.uk/data/Library2/pdf/1898-TS0016.pdf.

UK Treaties Online. "Convention between the United Kingdom and China respecting Tibet (To which is annexed the Convention between the United Kingdom and Tibet, signed at Lhasa, September 7, 1904)." http://foto.archivalware.co.uk/data/Library2/pdf/1906-TS0009.pdf.

UK Treaties Online. "Convention between the United Kingdom and China respecting WEI HAI WEI." http://foto.archivalware.co.uk/data/Library2/pdf/1898-TS0014.pdf.

UK Treaties Online. "Convention between the United Kingdom and Russia Relating to Persia, Afghanistan, and Tibet." http://foto.archivalware.co.uk/data/Library2/pdf/1907-TS0034.pdf.

UK Treaties Online. "Joint Declaration of the Government of the United Kingdom of Great Britain and Northern Ireland and the Government of the People's Republic of China on the Question of Hong Kong with Annexes." https://treaties.fco.gov.uk/awweb/pdfopener?md=1&did=68291.

UK Treaties Online. "Regulations respecting Trade in Tibet, concluded between the United Kingdom, China and Tibet." http://foto.archivalware.co.uk/data/Library2/pdf/1908-TS0035.pdf.

UK Treaties Online. "Treaty between His Majesty in respect of the United Kingdom and India and His Excellency the President of the National Government of the Republic of China for the Relinquishment of Extra-Territorial Rights in China and the Regulation of Related Matters (with Exchange of Notes and Agreed Minute)." https://treaties.fco.gov.uk/awweb/pdfopener?md=1&did=64833.

## Diaries and memoirs

Bai, Chongxi. *Bai Chongxi huiyilu [The Reminiscences of General Pai Chung-hsi]*. Beijing: Jiefangjun Chubanshe, 1987.

Duoji'ouzhu, Qiang'eba. "Xizang difangzhengfu pai 'daibiaotuan weiwen Tongmengguo he chuxi Nanjing Guomindaibiaodahui' neimu [Secret Story of Congratulations of the Tibetan Delegation to the Victory of the Allied Nations and Its Attendance to the Chinese National Assambly]." In *Xizang wenshiziliao xuanji (2) [Selected Collections of Historical Sources Relating to Tibet (Vol. 2)]*, edited by Xizang zizhiqu zhengxie wenshiziliaoyanjiuhui, 1–11. Beijing: Minzu Chubanshe, 1984.

Gao, Sulan, ed. *Jiang Zhongzheng zongtong dang'an: shilüe gaoben [Draft Papers of Chiang Kai-shek]*. 82 vols. Vol. 52–55. Taipei: Academia Historica, 2011.

Koo, Wellington. *Gu Weijue huiyilu [The Reminiscences of Dr. Wellington Koo]*. Translated by the Institute of Modern History at Chinese Academy of Social Sciences. 13 vols. Vol. 5. Beijing: Zhonghua Shuju, 2013.

Lin, May-li, ed. *Wang Shijie riji [The Diary of Dr. Wang Shih-chieh]*. 2 vols. Vol. 1. Taipei: Institute of Modern History Academia Sinica, 2012.

Liu, Shengqi. *Lasa jiushi [Memories regarding Lhasa]*. 2nd ed. Beijing: Zhongguo Zangxue Chubanshe, 2014.

Siwangduoji, Lalu. "Deli mimihuanwen weicengdedao yuan Xizang difangzhengfu chengren [The New Dehli Secret Exchanges of Notes Had Never Been Recognised by the Former Local Tibetan Government]." In *Xizang wenshiziliao xuanji (10) [Selected Collections of Historical Sources regarding Tibet (Vol. 10)]*, edited by Xizang zizhiqu zhengxie wenshiziliaoyanjiuhui, 6–13. Beijing: Minzu Chubanshe, 1989.

Wang, Zhenghua, ed. *Jiang Zhongzheng zongtong dang'an: shilüe gaoben [Draft Papers of Chiang Kai-shek]*. 82 vols. Vol. 60–62. Taipei: Academia Historica, 2011.

Wu, Peiying. "Jiang Zhongzheng zongtong dang'an: shilüe gaoben [The Historical Process of Chiang Kai-shek's Manoeuvre of Controlling Xikang on the Pretext of the Tibetan Affairs]." In *Wenshi Ziliao Xuanji (Vol. 33) [Selections of Historical Materials (Vol. 33)]*, edited by Zhengxie Quanguo Weiyuanhui, 140–154. Beijing: Wenshi Ziliao Chubanshe, 1985.

Xu, Jiatun. *Xu Jiatun huiyilu [The Memoir of Xu Jiatun]*. 2 vols. Vol. 2. Hong Kong: Lianjing Chunban Shiye Jituan, 1993.

Zeng, Sheng. *Zeng Sheng huiyilu [The Memoir of Zeng Sheng]*. Beijing: Jiefangjun Chubanshe, 1991.

Zhou, Meihua, ed. *Jiang Zhongzheng zongtong dang'an: shilüe gaoben [Draft Papers of Chiang Kai-shek]*. 82 vols. Vol. 48–51. Taipei: Academia Historica, 2011.

## Newspapers and journals

No Author. "American's Empire." *The New Statesman and Nation* 32, no. 822 (November 23, 1946): 371–372.

No Author. "Commercial Treaty Delay." *Shanghai Evening Post* 66, no. 155 (July 7, 1947).

No Author. "On Watch." *The Syren & Shipping Illustrated* 189, no. 2463 (November 10, 1943): 153–156.

Song, Meiling. "First Lady of the East Speaks to the West." *The New York Times*, April 19, 1942.

Times Correspondent. "American Treaty with China: Commerce and Navigation." *The Times*, November 5, 1946.

## Other published sources

Alexeeva, Olga. "Chinese Migration in the Russian Far East: A Historical and Sociodemographic Analysis." *China Perspectives* 75, no. 3 (2008): 20–32.

An, Haiyan. "Zuowei 'Zhuanlunwang' he 'Wenshupusa' de Qingdi: jianlun Qianlongdi yu Zangchuanfojiao de guanxi [Emperor as Chakravartin and the Incarnation of Mañjuśrī:

The Relationship between the Qianlong Emperor and Tibetan Buddhism]." *Qingshi Yanjiu [The Qing History Journal]*, no. 2 (2020): 105–118.

Baxter, Christopher. *The Great Power Struggle in East Asia, 1944–50: Britain, America and Post-war Rivalry*. London: Palgrave Macmillan, 2009.

Bell, Sir Charles. *Tibet: Past and Present*. Oxford: The Clarendon Press, 1924.

Bickers, Robert. *Britain in China: Community, Culture and Colonialism 1900–1949*. Manchester and New York: Manchester University Press, 1999.

Bickers, Robert. "Legal Fiction: Extraterritoriality as an Instrument of British Power in China in the 'Long Nineteenth Century'." In *Empire in Asia: A New Global History (The Long Nineteenth Century)*, 2 vols., Vol. 2, edited by Donna Brunero and Brian P. Farrell, 53–80. London: Bloomsbury Academic, 2018.

Bickers, Robert. *Out of China: How the Chinese Ended the Era of Western Domination*. London: Allen Lane – Penguin Random House, 2017.

Bickers, Robert. "Revisiting the Chinese Maritime Customs Service, 1854–1950." *The Journal of Imperial and Commonwealth History* 36, no. 2 (2008): 221–226.

Bickers, Robert. *The Scramble for China: Foreign Devils in the Qing Empire, 1832–1914*. London: Allen Lane, 2011.

Boecking, Felix. *No Great Wall: Trade, Tariffs, and Nationalism in Republican China, 1927–1945*. Cambridge (Massachusetts) and London: Harvard University Asia Center, 2017.

Brunero, Donna. *Britain's Imperial Cornerstone in China: the Chinese Maritime Customs Service, 1854–1949*. London and New York: Routledge, 2006.

Burns, John P. "Immigration from China and the Future of Hong Kong." *Asian Survey* 27, no. 6 (1987): 661–682.

Cao, Dachen. "Jindai Riben zai Hua lingshicaipanquan shulun [Review on Japanese Consular Jurisdiction in Modern China]." *Kangrizhanzheng yanjiu*, no. 1 (2008): 1–22.

Cassel, Pär Kristoffer. *Grounds of Judgement: Extraterritoriality and Imperial Power in Nineteenth-Century China and Japan*. Oxford and New York: Oxford University Press, 2012.

Ch'i, Hsi-sheng. "The Military Dimension, 1942–45." In *China's Bitter Victory: the War with Japan 1937–1945*, edited by James C. Hsiung and Steven I. Levine, 157–184. New York and London: M. E. Sharpe, 1992.

Ch'i, Hsi-sheng. *The Much Troubled Alliance: US-China Military Cooperation during the Pacific War, 1941–1945*. London: World Scientific Publishing Co., 2015.

Chan, Kit-ching Lau. "The Abrogation of British Extraterritoriality in China 1942–43: A Study of Anglo-American-Chinese Relations." *Modern Asian Studies* 11, no. 2 (1977): 257–291.

Chan, Kit-Ching Lau. *China, Britain and Hong Kong, 1895–1945*. Hong Kong: The Chinese University Press, 1990.

Chan, Kit-ching Lau. "Hong Kong in Sino-British Diplomacy, 1926–45." In *Precarious Balance: Hong Kong Between China and Britain 1842–1992*, edited by Ming K. Chan, 71–89. New York and London: M.E. Sharpe, 1994.

Chang, Chihyun. *Government, Imperialism and Nationalism in China: the Maritime Customs Service and Its Chinese Staff*. London and New York: Routledge, 2013.

Chen, Qianping. *Kangzhanqianhou zhi Zhong-Ying Xizang jiaoshe, 1935–1947 [Sino-British Discussions over Tibet before, during and after the Anti-Japanese War, 1935–1947]*. Beijing: Sanlianshudian, 2003.

Chen, Shiqi. *Zhongguo jindai haiguan shi [The Modern History of the Chinese Maritime Customs Service]*. Beijing: Renmin Chubanshe, 2002.

Chiang, Kai-shek. *China's Destiny and Chinese Economic Theory*. New York: Roy Publishers, 1947.

Clark, Douglas. *Gunboat Justice: British and American Law Courts in China and Japan, 1842–1943*. 3 vols. Vol. 1–3. Hong Kong: Earnshaw Books Limited, 2015.

Clifford, Nicholas. *Retreat from China: British Policy in the Far East, 1937–1941*. Seattle: University of Washington Press, 1967.

Cruickshank, Charles. *S.O.E. in the Far East*. Oxford and New York: Oxford University Press, 1983.

Fawcett, Brian C. "The Chinese Labour Corps in France 1917–1921." *Journal of the Hong Kong Branch of the Royal Asiatic Society* 40 (2000): 33–111.

Feng, Lin. "Zhanhou Zhong-Ying shangyue liuchan lunxi [The Study and Analaysis of the Abortion of the Sino-British Commercial Treaty after WWII]." In *The 2006 Youth Academic Symposium of the Institute of Modern History, the Chinese Academy of Social Sciences*, edited by the Institute of Modern History CASS, 494–518. Beijing: Social Sciences Academic Press, 2007.

Feng, Mingzhu. *Zhong-Ying Xizang jiaoshe yu Chuanzang bianqing, 1774–1925 [Negotiations between China and Britain in relation to Tibet and Situations on the Tibet-Sichuan Border, 1774–1925]*. Beijing: Zhongguo Zangxue Chubanshe, 2007.

Feng, Zhong-ping. *The British Government's China Policy, 1945–1950*. Keele: Keele University Press, 1994.

Fishel, Wesley R. *The End of Extraterritoriality in China*. Berkeley and Los Angeles: University of California Press, 1952.

Fitzsimons, M.A. *The Foreign Policy of the British Labour Government, 1945–1951*. Notre Dame, Indiana: University of Notre Dame Press, 1953.

Fox, Charles J. "The Sino-American Treaty – I." *Far Eastern Survey* 16, no. 12 (1947): 138–142.

Frankel, Joseph. *British Foreign Policy, 1945–1973*. London, New York and Toronto: Oxford University Press, 1975.

Friedman, Irving S. *British Relations with China, 1931–1939*. New York and London: Institute of Pacific Relations and Allen & Unwin, 1940.

Fung, Edmund S.K. "The Chinese Nationalists and the Unequal Treaties 1924–1931." *Modern Asian Studies* 21, no. 4 (1987): 793–819.

Fung, Edmund S.K. *The Diplomacy of Imperial Retreat: Britain's South China Policy, 1924–1931*. Hong Kong: Oxford University Press, 1991.

Goldstein, Melvyn C. *A History of Modern Tibet, 1913–1951: the Demise of the Lamaist State*. Berkeley, Los Angeles and London: University of California Press, 1991.

Goldstein, Melvyn C., Dawei Sherap, and William R. Siebenschuh. *A Tibetan Revolutionary: the Political Life and Times of Bapa Phüntso Wangye*. Berkeley, Los Angeles and London: University of California Press, 2004.

Guyot-Réchard, Bérénice. *Shadow States: India, China and the Himalayas, 1910–1962*. Cambridge: Cambridge University Press, 2017.

Hall, William Edward. *A Treatise on the Foreign Powers and Jurisdiction of the British Crown*. Oxford: Clarendon Press, 1894.

Han, Wing-Tak. "Bureaucracy and the Japanese Occupation of Hong Kong." In *Japan in Asia, 1942–1945*, edited by William H. Newell, 7–24. Singapore: Singapore University Press, 1981.

Hongwei, Zhou. *Yingguo, Eguo yu Zhongguo Xizang [Britain, Russia and Chinese Tibet]*. Beijing: Zhongguo Zangxue Chubanshe, 2000.

Horesh, Niv. *Shanghai's Bund and Beyond: British Banks, Banknote Issuance, and Monetary Policy in China, 1842–1937*. New Haven and London: Yale University Press, 2009.

Irwin, Douglas A., Petros C. Mavroidis, and Alan O. Sykes. *The Genesis of the GATT*. Cambridge and New York: Cambridge University Press, 2008.

Jones, Francis Clifford. *Extraterritoriality in Japan and the Diplomatic Relations Resulting in Its Abolition, 1853–1899*. New Haven: Yale University Press, 1931.

Kayaoglu, Turan. *Legal Imperialism: Sovereignty and Extraterritoriality in Japan, the Ottoman Empire, and China*. Cambridge: Cambridge University Press, 2010.

Keeton, G.W. *The Development of Extraterritoriality in China*. 2 vols. Vol. 1–2. London: Longmans, Green and Company, 1928.

Knaus, John Kenneth. *Beyond Shangri-la: America and Tibet's Move into the Twenty-First Century*. Durham, North Carolina: Duke University Press, 2012.

Knaus, John Kenneth. "Official Policies and Covert Programs: The U.S. State Department, the CIA, and the Tibetan Resistance." *Journal of Cold War Studies* 5, no. 3 (2003): 54–79.

Kobayashi, Ryosuke. "Tibet in the Era of 1911 Revolution." *Journal of Contemporary East Asia Studies* 3, no. 1 (2014): 91–113.

Koller, Christian. "The Recruitment of Colonial Troops in Africa and Asia and Their Deployment in Europe during the First World War." *Immigrants & Minorities* 26, no. 1–2 (2008): 111–133.

Lamb, Alastair. *British India and Tibet, 1766–1910*. 2nd ed. New York: Routledge & Kegan Paul, 1986.

Lamb, Alastair. *The China-India Border: the Origins of the Disputed Boundaries*. London, New York and Toronto: Oxford University Press, 1964.

Lamb, Alastair. *The McMahon Line (I): A Study in the Relations between India, China and Tibet, 1904 to 1914*. 2 vols. Vol. 1. London: Routledge & Kegan Paul; Toronto: University of Toronto Press, 1966.

Lamb, Alastair. *The McMahon Line (II): A Study in the Relations between India, China and Tibet, 1904 to 1914*. 2 vols. Vol. 2. London: Routledge & Kegan Paul; Toronto: University of Toronto Press, 1966.

Lamb, Alastair. *Tibet, China & India, 1914–1950: A History of Imperial Diplomacy*. Hertingfordbury: Roxford Books, 1989.

Lane, Kevin P. *Sovereignty and the Status Quo: the Historical Roots of China's Hong Kong Policy*. Boulder, San Francisco and Oxford: Westview Press, 1990.

Li, Shi'an. "The Extraterritoriality Negotiations of 1943 and the New Territories." *Modern Asian Studies* 30, no. 3 (1996): 617–650.

Li, Shi'an. *Taipingyangzhanzheng shiqi de Zhong-Ying guanxi [The Sino-British Relationship during the Pacific War]*. Beijing: China Social Sciences Press, 1994.

Li, Shi'an. *Zhanshi Yingguo dui Hua zhengce [Britain's China Policy during the Second World War]. Fanfaxisi zhanzheng shiqi de Zhongguo yu Shijie yanjiu [Research on China and the World during the Anti-Fascist War]*, 9 vols. Vol. 7, edited by Dekun Hu. Wuhan: Wuhan University Press, 2010.

Li, Yumin. "Wanqing Zhong-wai Bupingdengtiaoyue de falü guanxi [The Legal Connections among the Unequal Treaties of the Late Qing Era]." In *Jindai Zhong-wai guanxishi yanjiu [Study on the History of Modern Sino-Foreign Relations]*, edited by Junyi Zhang and Hongmin Chen, 1–23. Beijing: Social Sciences Academic Press, 2017.

Li, Yumin. *Zhongguo feiyue shi [The History of Abolishing Unequal Treaties in China]*. Beijing: Zhonghua Shuju, 2005.

Lieu, D.K. *Foreign Investments in China*. Nanjing: Chinese Government Bureau of Statistics, 1929.

Lin, Hsiao-ting. *Taihai, Lengzhan, Chiang Kai-shek: jiemi dang'an zhong xiaoshi de Taiwan shi 1949–1988 [Taiwan Strait, Cold War and Chiang Kai-shek: the Re-discovered Taiwan History in the Unclassified Archives 1949–1988]*. Taipei: Lianjing Chuban Shiye Gufen Youxian Gongsi, 2015.

Lin, Hsiao-ting. *Tibet and Nationalist China's Frontier: Intrigues and Ethnopolitics, 1928–49*. Vancouver and Toronto: University of British Columbia Press, 2006.

Liu, Xiaoyuan. *A Partnership for Disorder: China, the United States, and Their Policies for the Postwar Disposition of the Japanese Empire, 1941–1945*. New York: Cambridge University Press, 1996.

Loh, Christine. *Underground Front: the Chinese Communist Party in Hong Kong*. Hong Kong University Press, 2010.

Louis, William Roger. *Imperialism at Bay: the United States and Decolonization of the British Empire, 1941–1945*. New York: Oxford University Press, 1977.

Louis, William Roger. "Introduction." In *The Oxford History of the British Empire: the Twentieth Century*, 5 vols. Vol. 4, edited by Judith M. Brown and William Roger Louis, 1–46. Oxford: Oxford University Press, 1999.

Louis, William Roger. "Foreword." in *The Oxford History of the British Empire: Historiography*, 5 vols. Vol. 5, edited by Robin W. Winks, vii–xi. Oxford: Oxford University Press, 1999.

Lowe, Peter. *Britain in the Far East: A Survey from 1819 to the Present*. London: Longman, 1981.

Mark, Chi-Kwan. *The Everyday Cold War: Britain and China, 1950–1972*. London: Bloomsbury Academic, 2017.

Mark, Chi-Kwan. "The 'Problem of People': British Colonials, Cold War Powers, and the Chinese Refugees in Hong Kong, 1949–62." *Modern Asian Studies* 41, no. 6 (2007): 1145–1181.

McKay, Alex. *Tibet and the British Raj: the Frontier Cadre 1904–1947*. Richmond: Curzon Press, 1997.

Mitter, Rana. *A Bitter Revolution: China's Struggle with the Modern World*. New York: Oxford University Press, 2005.

Mitter, Rana. *China's War with Japan 1937–1945: the Struggle for Survival*. London: Penguin Books, 2013.

Mullaney, Thomas S. *Coming to Terms with the Nation: Ethnic Classification in Modern China*. Berkeley, Los Angeles and London: University of California Press, 2011.

Nield, Robert. "Treaty Ports and Other Foreign Stations in China." *Journal of the Royal Asiatic Society Hong Kong Branch* 50 (2010): 123–129.

Osterhammel, Jürgen. "Britain and China, 1842–1914." In *The Oxford History of the British Empire: the Nineteenth Century*, 5 vols. Vol. 3, edited by Andrew Porter, 146–169. Oxford: The Oxford University Press, 1999.

Osterhammel, Jürgen. "China." In *The Oxford History of the British Empire: the Twentieth Century*, 5 vols. Vol. 4, edited by Judith M. Brown and William Roger Louis, 643–666. Oxford: Oxford University Press, 1999.

Ovendale, Ritchie. "Introduction." In *The Foreign Policy of the British Labour Governments, 1945–1951*, edited by Ritchie Ovendale, 1–20. Leicester: Leicester University Press, 1984.

Pepper, Suzanne. *Civil War in China: the Political Struggle, 1945–1949.* Lanham, Md.; Oxford: Roman & Littlefield Publishers, 1999.

Petech, Luciano. *China and Tibet in the Early 18th Century: History of the Establishment of Chinese Protectorate in Tibet.* Leiden: E. J. Brill, 1950.

Piggott, Francis. *Exterritoriality: the Law Relating to Consular Jurisdiction and to Residence in Oriental Countries.* London: William Clowes and Sons Limited, 1892.

Plating, John D. *The Hump: America's Strategy for Keeping China in World War II.* College Station: Texas A&M University Press, 2011.

Porter, Brian. *Britain and the Rise of Communist China: A Study of British Attitudes, 1945–1954.* London and Toronto: Oxford University Press, 1967.

Reid, Robert. *History of Frontier Areas Bordering on Assam 1883–1941.* Delhi: Eastern Publishing House, 1983.

Relyea, Scott. "Indigenizing International Law in Early Twentieth-Century China: Territorial Sovereignty in the Sino-Tibetan Borderland." *Late Imperial China* 38, no. 2 (December 2017): 1–60.

Remer, C.F. *Foreign Investments in China.* New York: H. Fertig, 1933.

Ride, Edwin. *B.A.A.G., Hong Kong Resistance, 1942–1945.* Hong Kong: Oxford University Press, 1981.

Roosevelt, Elliott. *As He Saw It.* New York: Duell, Sloan, and Pearce, 1946.

Saville, John. *The Politics of Continuity: British Foreign Policy and the Labour Government, 1945–46.* London and New York: Verso, 1993.

Schake, Kori. *Safe Passage: the Transition from British to American Hegemony.* Cambridge, Massachusetts: Harvard University Press, 2017.

Schenk, Catherine R. "Commercial Rivalry between Shanghai and Hong Kong during the Collapse of the Nationalist Regime in China, 1945–1949." *The International History Review* 20, no. 1 (1998): 68–88.

Schenk, Catherine R. *Hong Kong as an International Financial Centre: Emergence and Development 1945–65.* London and New York: Routledge, 2001.

Shai, Aron. *Britain and China, 1941–47: Imperial Momentum.* London: Macmillan Press in association with St. Antony's College, Oxford, 1984.

Shi, Zhe. *Zai lishi juren shenbian [Standing Beside Giant Man in History].* Beijing: Zhongyangwenxian Chubanshe, 1991.

Singh, Amar Kaur Jasbir. *Himalayan Triangle: A Historical Survey of British India's Relations with Tibet, Sikkim and Bhutan 1765–1950.* London: The British Library, 1988.

Smith Jr., Warren W. *Tibetan Nation: A History of Tibetan Nationalism and Sino-Tibetan Relations.* Boulder, Colorado and Oxford: Westview Press, 1996.

Stephens, Thomas B. *Order and Discipline in China: the Shanghai Mixed Court 1911–27.* Seattle and London: University of Washington Press, 1992.

Sun, Yang. *Wuguo'erzhong: Zhanhou Zhong-Ying Xianggang wenti jiaoshe, 1945–1949 [An Ending Without Result: Sino-British Post-war Discussions over the Matter of Hong Kong, 1945–1949]*. Beijing: Social Sciences Academic Press, 2014.

Tang, James T.H. "World War to Cold War: Hong Kong's Future and Anglo-Chinese Interactions, 1941–55." In *Precarious Balance: Hong Kong between China and Britain, 1842–1992*, edited by Ming K. Chan, 107–130. New York: M.E. Sharpe, 1994.

Tang, James Tuck-Hong. *Britain's Encounter with Revolutionary China, 1949–54*. London and New York: St. Martin's Press, 1992.

Tarring, Charles James. *British Consular Jurisdiction in the East: With Topical Indices of Cases on Appeal from and Relating to Consular Courts and Consuls, also a Collection of Statutes Concerning Consuls*. London: Stevens and Haynes, 1887.

Thorne, Christopher. *Allies of a Kind: the United States, Britain, and the War against Japan, 1941–1945*. New York: Oxford University Press, 1978.

Tsang, Steve. *Governing Hong Kong: Administrative Officers from the Nineteenth Century to the Handover to China, 1862–1997*. London and New York: I. B. Tauris & Co Ltd., 2007.

Tsang, Steve. *Hong Kong: An Appointment with China*. London and New York: I. B. Tauris & Co Ltd., 1997.

Tsang, Steve. *A Modern History of Hong Kong*. London and New York: I. B. Tauris & Co Ltd., 2004.

Tuttle, Gray. *Tibetan Buddhists in the Making of Modern China*. New York: Columbia University Press, 2005.

van de Ven, Hans. *Breaking with the Past: the Maritime Customs Service and the Global Origins of Modernity in China*. New York: Columbia University Press, 2014.

van de Ven, Hans. *China at War: Triumph and Tragedy in the Emergence of the New China, 1937–1952*. London: Profile Books, 2017.

van de Ven, Hans. "The Sino-Japanese War in History." In *The Battle for China: Essays on the Military History of the Sino-Japanese War of 1937–1945*, edited by Mark Peattie, Edward J. Drea and Hans van de Ven, 446–466. Stanford, California: Stanford University Press, 2011.

van de Ven, Hans. *War and Nationalism in China, 1925–1945*. London and New York: Routledge Curzon, 2003.

van Walt van Praag, Michael C. *The Status of Tibet: History, Rights, and Prospects in International Law*. Boulder, Colorado: Westview Press, 1987.

Wang, Gengxiong, ed. *Sun Zongshan ji wai ji [The Supplementary Anthology of Dr. Sun Yat-sen]*. Shanghai: Shanghai Renmin Chubanshe, 1990.

Wang, Jianlang. *Zhongguo feichu Bupingdeng tiaoyue de licheng [The Historical Process of Chinese Abolition of Unequal Treaties]*. Nanchang: Jiangxi Renmin Chubanshe, 2000.

Wei, C.X. George. *Sino-American Economic Relations, 1944–1949*. Westport, Connecticut and London: Greenwood Press, 1997.

Wesley-Smith, Peter. *Unequal Treaty 1898–1997: China, Great Britain and Hong Kong's New Territories*. Hong Kong: Oxford University Press, 1980.

Whitfield, Andrew J. *Hong Kong, Empire and the Anglo-American Alliance at War, 1941–45*. Basingstoke and New York: Palgrave, 2001.

Wu, Qine. "Zhonghuaminzu zongzulun yu Zhonghuaminguo de bianjiangzizhi shijian [The Theory of the Clan of Chinese Nation and the Practice of Local Autonomy in the Frontier of the Republic of China]." In *Chiang Kai-shek and the Formation of Modern China*, 2 vols. Vol. 1, edited by Zijin Huang and Guangzhe Pan, 162–213. Taipei: Institute of Modern History, Academia Sinica, 2013.

Xiang, Lanxin. *Recasting the Imperial Far East: Britain and America in China, 1945–1950.* New York: M.E. Sharpe, 1995.

Yu, Maochun. "'In God We Trusted, In China We Busted': the China Commando Group of the Special Operations Executive (SOE)." *Intelligence and National Security* 16, no. 4 (2001): 37–60.

Zhai, Qiang. *The Dragon, the Lion, and the Eagle: Chinese-British-American Relations, 1949–1958.* Kent, Ohio and London, England: The Kent State University Press, 1994.

Zhai, Qiang. "Tibet and Chinese-British-American Relations in the Early 1950s." *Journal of Cold War Studies* 8, no. 3 (2006): 34–53.

Zhang, Hao. "1946 nian Xizang difangdaibiao chuxi zhixianguoda wenti zhi tanxi [The Analysis of the Tibetan Attendance at the National Assembly in 1946]." *Shixue Jikan* 4 (July 2013): 64–79.

Zhang, Hao. "Zhanhou de Xizang wenti yu 'Zhonghuaminguo xianfa' youguan tiaowen de zhiding ji shishi [The Post-war Tibet Question and the Stipulation and Implementation of the Relevant Clauses of the 1946 Constitution of the Republic of China]." *Shixue Yuekan* 2 (2015): 51–65.

Zhang, Junyi. "1948 nian Guangzhou Shamianshijian shimo: yi Song Ziwen dang'an wei zhongxin [The History of the 1948 Shamian Incident in Guangzhou: A Study Based on the T.V. Soong Papers]." *Social Sciences in China*, no. 6 (2008): 185–200.

Zhu, Lishuang. *Minguozhengfu de Xizang zhuanshi (1912–1949) [Envoys of the Republic Governments on Their Missions to Tibet, 1912–1949].* Hong Kong: The Chinese University Press, 2016.

# Index

https://doi.org/10.1515/9783110706659-016